D0123170

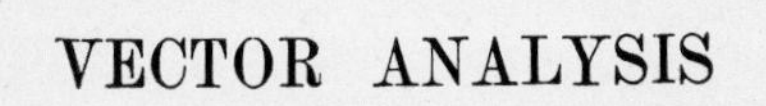

VECTOR ANALYSIS

The author will be grateful to any reader who will call his attention to an error in this book, or who will suggest new problems, changes in the text, or additions thereto. Careful consideration will be given to such suggestions and criticisms, as it is the author's desire to make his book as nearly perfect as possible. Communications of this nature should be addressed to Dr. J. G. Coffin, 199 Elizabeth Ave., Hempstead, N. Y.

VECTOR ANALYSIS

AN INTRODUCTION
TO
VECTOR-METHODS
AND THEIR VARIOUS APPLICATIONS
TO
PHYSICS AND MATHEMATICS

BY

JOSEPH GEORGE COFFIN, B.S., PH.D.

(MASS. INST. TECH. '98 AND CLARK UNIVERSITY '03)
EX-ASSOCIATE PROFESSOR OF PHYSICS AT THE COLLEGE OF THE
CITY OF NEW YORK
CONSULTING AND DEVELOPMENT ENGINEER WITH THE
GENERAL BAKING COMPANY

SECOND EDITION

NEW YORK
JOHN WILEY & SONS, INC.
LONDON: CHAPMAN & HALL, LIMITED

Printed in U. S. A.

Stanhope Press
F. H. GILSON COMPANY
BOSTON, U.S.A.

3-30

PREFACE

Ever since the development of Quaternion analysis, by Sir William Rowan Hamilton, and of the "Ausdehnungslehre," by Grassmann, there has been a growing feeling that the older and more common processes of analysis were in some way artificial and complex.

This fact exists, for it is such, because these newer methods and ideas apply more naturally, more simply and more directly to many of the conceptions of geometry, mechanics and mathematical physics, than those long accepted.

Why then have these admitted advantages not led to a more universal adoption of these methods? The answer seems to be that the required change of ideas, of manner of thought and of notation, was too radical. It is well known that changes evolve slowly, and although to many, evolution is far too slow a process, the only way to proceed is to aid to the best of one's ability in bringing about the desired result.

One who has studied and labored over the applications of mathematical analysis to physical and geometrical problems, naturally has reluctance to discard the old familiar looking formulæ and start anew in an unknown and radically different language.

However great the skill and ingenuity shown by the pioneer in solving problems by Quaternions, there was always left the thought to the unbiased student that a lack of parallelism existed between the old and the new methods of treatment. Such a lack undoubtedly does exist, but it is only during the last few years that a method has been evolved which avoids this fatal defect. It is chiefly through the labors of Gibbs and Heaviside that an analysis has been perfected which not only does away with the *unnecessary* complexity and artificiality of other analyses but offers a strictly natural and therefore as direct and simple a substitute as possible, and, at the same time in no wise is at variance, but runs parallel to them.

This new, yet old method is Vector Analysis; it combines within itself most of the advantages of both Quaternions and of Cartesian Analysis.

The adoption of Vector Analysis is urged on the grounds of naturalness, simplicity and directness; with it the true meaning of processes and results is brought out as clearly as possible, and desirable abbreviation is obtained.

It is admitted, that to a straight and clear thinker, almost any notation or mathematical method suffices, and to such a one, changes in notation or method may appear hardly worth while. He has already attained one of the results which, perforce, follow the intelligent assimilation of a vector method of thinking. To him there is left but the attainment of a simple notation which is the logical accompaniment of clear thought. A few examples of vector concentration are to be found in the exercises of the last chapter of this book. But the sole use of vector notation, without the insight and clear conceptions which should obtain at the same time, is without any value whatsoever, vitiates the vector point of view, and is contrary to the spirit of it.

It is almost unnecessary to state that the mind of the physicist ought to be of the visual type so well exemplified in the mind of Faraday. He should see the lines of force emerging from the magnet; see that they are continuous within the metal; follow them, in his mind's eye, as they are displaced by various causes; he should have some sort of a visual conception of the manner in which the electro-magnetic waves are traveling through the ether around him; to him the divergence and the divergence theorem should have a simple meaning.

To a mind other than this, the study of mathematical physics must be merely a series of analytical transformations without the vitality of their visual significance. To purely analytical minds, as distinct from the visual or intuitive type, the methods of Vector Analysis reduce to little more than an analytical shorthand. To the intuitive mind, however, they are illuminative and simplifying, allowing the mind to grasp and the hand to write the essential facts and transformations, unembarrassed with the generally undesirable complexity of Cartesian symbolism. It is impossible to study and to apply Vector Analysis to problems and not to have one's ideas and thought made clearer and better by the labor involved.

There are very good reasons for all these advantages. In Nature we are confronted with quantities called scalars which have size or magnitude only, and also with other quantities called vectors which have direction as well as magnitude. In order to manipulate vector quantities by the older methods, they were decomposed into three components along three arbitrary axes and the operations made upon these components. Is it not evident that the bringing in of three arbitrary axes is an artificial process, and that the decomposition of the vector into components along these axes is also artificial, unnatural even? Why not go directly to the vector itself and manipulate it without axes and without components? To do this, is possible, and in the following pages an attempt is made to show how it may be done.

There is still another ground for urging more extended study of Vector Analysis than now obtains. So many physicists of renown have been converted to its methods and use that to ignore their leadership is an impossibility. When such men as Lorentz, Föppl, Heaviside, Bucherer, Gibbs,

Abraham, Bjerknes, Sommerfeld, Cohn and many others are converted to its use, it is high time that the student familiarize himself at least with vector notation, even if not to become an expert in its use.

No one can deny the vast improvement that has taken place, in recent years, in our conceptions of physical processes; and few will deny that a large part of this improvement has been due to the ideas introduced with the advent of vector methods of thought.

That Lagrange reduced all of mechanics to a purely analytical basis without, as he boasts, necessitating diagrams, is certainly a wonderful accomplishment. Yet how much clearer and more elegant if the equations become alive with meaning, if to the algebraic transformations a mental picture of what is taking place is obtained!

Maxwell gave a splendid reference in favor of the new methods when he said, in speaking about the motion of the top, "Poinsot has brought the subject under the power of a more searching analysis than that of the calculus, in which ideas take the place of symbols and intelligible propositions supersede equations."

Vector Analysis has the advantages of Lagrange's analytical method as well as those of the idealogical method of Poinsot.

The writer does not, in any way, urge the rejection of anything of value in any method whatsoever. It is not well nor is it intended that the methods of Vector Analysis should be essentially different from those to which the student is supposed to be accustomed. In fact, it has been the aim throughout this book to evolve an analysis to which all the knowledge of the reader can be immediately applied, and to so expound this analysis, that Cartesian equations may be immediately written in vector notation and conversely.

There is still another important advantage, which should not be overlooked, that is, vector notation just as vector thought, is entirely independent of any choice of axes, or planes of reference, and yet the transformation of the vector equations into other systems, requiring these axes or planes is always extremely easy. To prove that a *natural* invariant is invariant to a change of axes, has always appeared to the writer an extremely foolish operation and a waste of time. This is not saying, that in a mathematical theory of invariants such a property of an algebraic expression is not instructive or interesting. But to say, for example, that the properties of the lines of force which cut a set of equipotential surfaces at right angles, (*i.e.*, the lines $\mathbf{F} = - \nabla V$) may be dependent upon the particular set of axes used to investigate them, is a waste of time to say the least. How can a truth vary with the language used in expressing it?

No attempt at mathematical rigor is made. Such refinements serve only to conceal the simplicity of fact, which it is the aim of these pages to elucidate. The appearance of extended proofs, the writer considers to be entirely out of place in a book of this kind. On the other hand, no one is more in favor of mathematical rigor than he; the point is simply to eliminate discussions whose presence would lead the attention astray from the main ideas of the argument. In any case, whenever a demonstration does not satisfy the fastidious, the results may be found more rigorously, if not more clearly, established in works devoted to mathematically rigorous demonstrations.

The student will find with a little study that he may easily take down lectures, given in Cartesian notation, directly into vector notation. Serious trial will convince him that time is gained and what is still more important, that equations will be, *must* be, understood if this is done. It is by precisely such a process that the writer familiarized himself with the subject.

The notation adopted is that of Prof. Willard Gibbs, one of the too few great American physicists and mathematicians. The reasons leading to this choice are fully set forth in the Appendix.

The first part of the book is devoted to a concise treatment of the fundamental principles of the subject, the remaining chapters, to the application of the analysis to the *beginnings* of mathematical physics, including geometry, mechanics, magnetism, electricity, heat and hydrodynamics. It was found necessary to omit many beautiful applications in elasticity, electron theory and other parts of physics in order to keep the size of the volume within bounds.

The student who takes up the later chapters, is supposed to be familiar, to a certain extent, with the subjects therein contained, and these chapters are intended to show the beginner how to translate and demonstrate the theorems into the new calculus. The writer therefore makes this his apology for a certain necessary lack of logical sequence in the treatment of the various subjects.

The treatment of alternating currents and allied subjects has been omitted, because in practically every modern book on the subject the notation of the special vector method employed, is fully explained in some part of it.

It is hoped that but few errors still remain in the text. The author, alone, corrected the proof, but numerous equations and special difficulties met with in printing in a new notation, rendered the corrections very difficult and laborious.

The copy has been read by Prof. Saurel, professor of mathematics in the College of the City of New York, and the author wishes to acknowledge here his indebtedness for the kindness as well as for many valuable suggestions.

A detailed list of works on Quaternions is rendered unnecessary by Professor Macfarlane's "Bibliography" published by the "Association for the Promotion of the Study of Quaternions and Allied Mathematics," Dublin, 1904, but a list of works which have been especially consulted is appended to the preface, and the writer here acknowledges his obligations to all of them. If this book succeeds in making plain the author's particular point of view; in simplifying ideas, or in causing simple ideas to seem clearer than before, he will feel amply repaid for any pains taken in producing what was to him a labor of love.

J. G. COFFIN.

NEW YORK, April 9, 1909.

PREFACE TO SECOND EDITION

IN this new edition a number of small errors which are peculiarly difficult of discovery in a work involving so many different kinds of type have been corrected. The sincere thanks of the writer are due to the large number of correspondents who have greatly helped him in this revision.

The author is glad to be able to state that to his knowledge but one theoretical error has been discovered up to the present time.

Certain portions have been rewritten and fourteen pages of notes have been added to the appendix.

In particular a short digression on different varieties of vectors; certain additional definitions of differential geometry with reference to curves in space which seemed interesting and useful; the demonstration of Frenet's valuable formulæ for space curves; an interesting example of vector reasoning as applied to the solution of the differential equation of motion of an electron in a magnetic field; two new proofs of Stokes' Theorem not found as far as we know in any treatise of vector analysis; an additional proof of Gauss's Theorem; and proofs of two theorems in integration analogous to the Divergence Theorem.

Both the publisher and the writer are delighted with the reception accorded this little book in this country and abroad.

The writer is of the opinion that a great many results of mathematical physics are elementary and easily understood by the student if explained in the right way, and the student thereby finds himself in a position to go right ahead in the more difficult extensions, when he comes to them. This book was written with that end in view. It is practically an elementary course in mathematical physics.

He also hopes that not only will this volume help the student to an acquisition of the fundamentals of Vector Analysis, but that also, and not least, it will awaken in him a desire for further study in that most beautiful and extensive of all branches of study, — Mathematical Physics.

He believes that in this country there is a wealth of material for the making of brilliant investigators in this line, if they are encouraged to approach the higher branches without the fear that it is beyond their capabilities.

He therefore makes a plea for the encouragement of students having ability in this direction so that soon it can no longer be said that we are not up to the standard of the investigators of the Old World. True, they had a long start and we have been handicapped, but we hope in the course of a few years to be abreast of them.

J. G. Coffin.

New York, *June*, 1911.

CONTENTS.

CHAPTER I.

Elementary Operations of Vector Analysis.

CHAPTER II.

Scalar and Vector Products of Two Vectors.

CHAPTER III.

Vector and Scalar Products of Three Vectors.

CHAPTER IV.

Differentiation of Vectors.

CHAPTER V.

THE DIFFERENTIAL OPERATORS.

$$\nabla \equiv \mathbf{i}\frac{\partial}{\partial x} + \mathbf{j}\frac{\partial}{\partial y} + \mathbf{k}\frac{\partial}{\partial z}.$$

CHAPTER VI.

APPLICATIONS TO ELECTRICAL THEORY.

CHAPTER VII.

Applications to Dynamics, Mechanics and Hydrodynamics.

CONTENTS.

APPENDIX.

NOTATION AND FORMULÆ.

FORMULÆ.

BIBLIOGRAPHY

Works Specially Consulted in the Preparation of this Book.

APPELL. Traité de Mécanique Rationelle.

BJERKNES. Vorlesungen über Hydrodynamischen Fernkräfte.

BUCHERER. Elemente der Vektor-Analysis.

BURNSIDE and PANTON. Theory of Equations.

CLIFFORD. Elements of Dynamic.

DRUDE. Theory of Optics.

EMTAGE. Introduction to the Mathematical Theory of Electricity and Magnetism.

FEHR. Méthode Vectorielle de Grassmann.

FISCHER. Vektordifferentiation und Vektorintegration.

FÖPPL. Maxwell'sche Theorie der Elektricität.

——. Vorlesungen über Technische Mechanik.

GANS. Einführung in die Vektoranalysis.

GIBBS. Collected Papers.

HEAVISIDE. Electrical Papers.

——. Electro-magnetic Theory.

HENRICI and TURNER. Vectors and Rotors with Applications.

IBBETSON. Mathematical Theory of Elasticity.

JAUMANN. Bewegungslehre.

JOLY. Manual of Quaternions.

KELLAND and TAIT. Introduction to Quaternions.

KIRCHHOFF. Vorlesungen über Mathematische Physik.

LAGRANGE. Mécanique Analytique.

LOVE. Theory of Elasticity.

MAXWELL. Electricity and Magnetism.

MCAULAY. Utility of Quaternions in Physics.

——. Octonions.

MINCHIN. Treatise on Statics.

PIERCE, B. O. Elements of the Theory of the Newtonian Potential Function.

POINSOT. Théorie Nouvelle de la Rotation des Corps.

ROUTH. Rigid Dynamics.

STEINMETZ. Alternating Current Phenomena.

TAIT. Dynamics.

——. Quaternions.

WALTON. Problems in Mechanics.

WEBSTER. The Dynamics of a Particle and of Rigid, Elastic and Fluid Bodies.

——. The Theory of Electricity and Magnetism.

WILLIAMSON and TARLETON. Elementary Treatise on Dynamics.

WILSON, GIBBS-. Vector Analysis.

SUGGESTIONS FOR WRITING VECTOR ANALYSIS ON THE BOARD

A number of inquiries have come in asking how to write vectors on the blackboard. It seems that the bold-faced type or **Clarendon** is perfectly satisfactory as far as print is concerned, but it is impracticable to produce such a difference in chalk-written symbols. To a great extent these same troubles also occur in manuscript.

There are several methods of differentiating vectors from purely scalar symbols which have proved satisfactory.

The notation given in the text is entirely practicable and definite. That is, if a, b or r denote vectors in any discussion, let a_0, b_0 or r_0 denote their magnitudes and a_1, b_1 or r_1 denote their directions or unit vectors along a, b or r respectively.

Thus $$a = a_0 a_1.$$

There is here a slight chance of ambiguity in the equation

$$a = a_1 i + a_2 j + a_3 k$$

where the i-component of a might be confounded with the unit vector along a. The writer does not consider this a serious objection. Like Tait we say that anybody finding difficulty with this small matter has begun the study of vectors too soon!

Another method is to place a line or dash over the vector-symbol. So that if $\bar{a}$ denotes a vector, then a is its magnitude, $\bar{a}_1$ is its direction and a_1 is its i-component.

Still another method to which the writer is very partial, having been brought up on Hamilton's notation, is to reserve the Greek alphabet for vectors.

So that if α is any vector, α_0 or a is its magnitude, and α_1 is its direction, while a_1, a_2, a_3 are its i-, j-, k-components.

After reading the book notices and reviews we are still of the same opinion as to the essential superiority of Gibb's notation over others, notwithstanding the criticisms of it which we expected. We have never claimed that $a \times b$ was a symmetrical *function,* but we do claim that the *notation* of both $a \cdot b$ and $a \times b$ is symmetrical. The minus sign in $a \times b = - b \times a$, does not make the *notation* unsymmetrical.

Almost simultaneously with this text was issued a vector analysis by the Italian mathematicians Burali-Forti and Marcolongo.

These gentlemen have invented still another notation which is similar to ours but which employs the $\times$ (large cross) for a scalar product and an inverted V (Λ) for a vector product. With the symmetry of their notation we are in favor, but why introduce any more notations when there are already so many to pick from?

This question of notation, which has nothing to do with the spirit of the method, is for each individual to solve for himself. We have employed what *we* believe to be the simplest and best and we have presented at length our arguments in favor of it.

J. G. COFFIN, 1911

VECTOR ANALYSIS

CHAPTER I.

ELEMENTARY OPERATIONS OF VECTOR ANALYSIS.

Definitions.

1. A Vector is a directed segment of a straight line on which are distinguished an initial and a terminal point. A vector thus has a magnitude and a direction. Any quantity which can be represented by such a segment may be called a vector quantity. The importance of this generalized conception is easily understood when it is considered that motion or displacement, velocity, acceleration, force, electric current, magnetic flux, lines of force, stresses and strains due to any cause, flow of heat and of fluids, all involve two parts, *i.e.*, magnitude and direction. All such quantities are vector quantities.

A Scalar on the other hand is any quantity which although having magnitude does not involve direction. For example, mass, density, temperature, energy, quantity of heat, electric charge, potential, ocean depths, rainfall, numerical statistics such as birth rates, mortality or population, are all scalar quantities.

A scalar, then, reduced to its simplest terms is merely a number and as such obeys all the laws of ordinary algebraic analysis. A vector, however, involving direction in addition to its numerical magnitude has an analysis peculiar to itself, the laws of which are to be derived.

2. Graphical Representation of a Vector. Any vector quantity may be represented graphically by an arrow.

The tail of the arrow, O, is called the origin; the head, A, is called the end or terminus.

Symbolically a vector may be denoted by two letters, the first one indicating the origin, the second one the terminus.

A small arrow is often placed over these letters to indicate more exactly that the quantity considered is a vector. Thus, $\overrightarrow{OA}$ denotes the vector beginning at O, ending at A, and pointing in the direction from O to A. This notation while useful is at times cumbersome. Hence more usually a vector will be denoted by a single letter, which involving more than a mere scalar is printed differently to distinguish it from purely scalar quantities, *i.e.*, **in Bold-faced Type.**

Thus the vector **a*** means the going of the distance OA in the direction O to A from *any point in space as origin.*

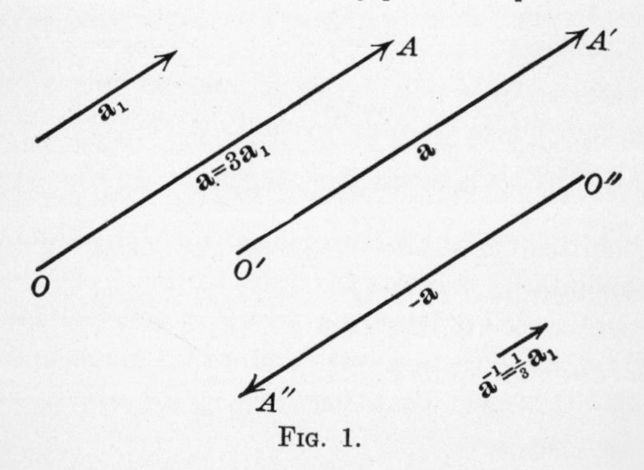

FIG. 1.

3. Equality of Vectors. All lines having the same length or magnitude and the same **sense** are equal vectors whatever their origin may be. Thus in Fig. 1, OA and $O'A'$ are equal vectors.

Negative Vector. The vector $O''A''$ having the same length and direction as **a** but the opposite sense, is defined

* The terms Step, Stroke, or Directed Magnitude are sometimes used as synonyms of Vector.

as the negative of **a** and is written $-\mathbf{a}$. Evidently also OA is the negative of $O''A''$.

Unit Vector. The directional part of any vector **a** may be concisely represented by a vector having the same sense and direction as **a** but of unit length. Such a vector is called a unit vector and will be denoted by adding the suffix 1 to the symbol representing the vector. Thus $\mathbf{a}_1$ is a vector having the same direction as **a**, but of unit length.

The length of a vector is termed its magnitude, size, or its absolute value. Sometimes, also, the term tensor is used. The magnitude of a vector **a** will be written a, using the same letter as that which denotes the vector but printed in *italic* type. It will be sometimes convenient also to denote the magnitude of **a** by adding the subscript 0 to **a** thus:

$$\mathbf{a}_0 \equiv a.$$

The vector **a** then may be considered as one, a times as long as $\mathbf{a}_1$ and hence we may write:

$$\mathbf{a} = a\,\mathbf{a}_1 \text{ or } = \mathbf{a}_0\mathbf{a}_1. \tag{1}$$

Any vector then may be represented by the product of its unit vector into its magnitude as in (1).

The expression $m\,\mathbf{a}$ denotes a vector m times as long as **a**, having the same direction but m times its magnitude. The multiplier -1 from what has been said about negative vectors, reverses a vector.

Parallel vectors whatever their magnitude are said to be *collinear*.

Reciprocal Vector. The vector parallel to **a** but whose length is the reciprocal of the length of **a** is said to be the reciprocal of **a**.

So that if

$$\mathbf{a} = a\,\mathbf{a}_1$$

$$\frac{1}{\mathbf{a}} = \mathbf{a}^{-1} = \frac{\mathbf{a}_1}{a}. \tag{2}$$

Composition of Vectors.*

4. Addition and Subtraction. To obtain graphically the sum of the two vectors **a** and **b**, draw **b** starting from the end of **a**; the line joining the origin of **a** with the end of **b** is the sum in question. In other words, it is the diagonal of the parallelogram of which the two vectors **a** and **b** are the sides.

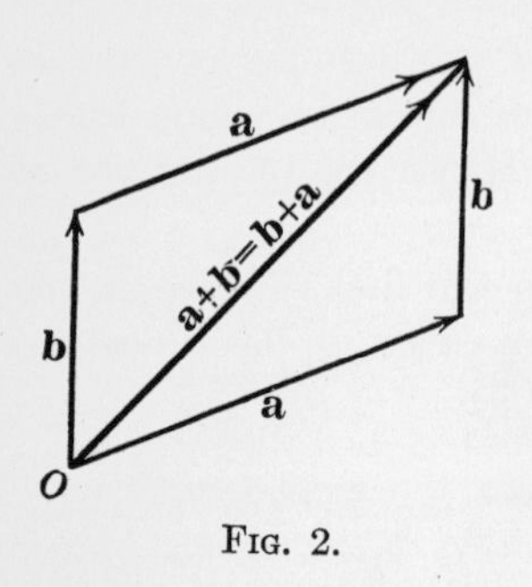

FIG. 2.

Evidently the sum (**a** + **b**) is the same as (**b** + **a**). If there are more than two vectors to be added, the sum of the first two may be taken and the third added to it as above, then to the resultant add the next one and so on. A moment's consideration of Fig. 3 will show that

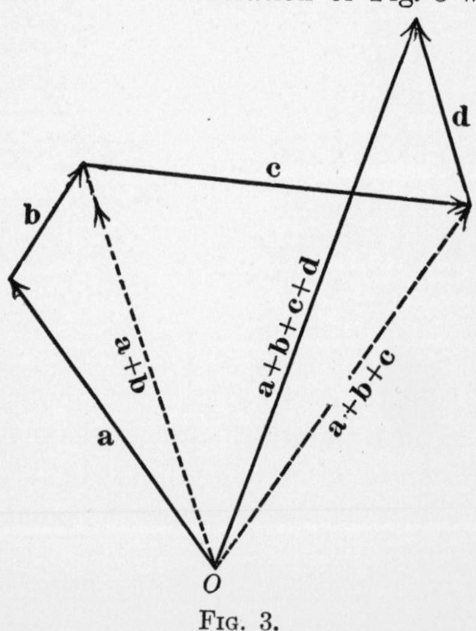

FIG. 3.

if we draw the vectors one after the other in a chain, each new one from the end of the last one drawn, the line joining

* See Appendix, p. 240, Note on Different Varieties of Vectors.

the origin of the first one to the terminus of the last one is the vector sum of them all.

A consideration of Fig. 3 will also show that the order in which they are taken is immaterial. The same construction then is used to find the sum of any number of vectors as is used in finding the resultant of the forces which would be represented by these vectors. Hence the importance of vectors in *mechanics*.

To subtract two vectors, add to the first the second one reversed. The extension of these rules to both positive and negative is obvious.

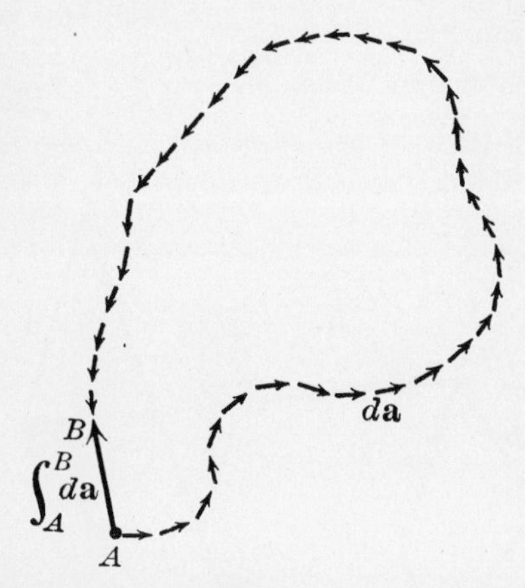

FIG. 4.

Vector Sum as an Integration. Any curve may be considered to be built up of an infinite number of infinitely short vectors, their directions being at every point along the tangent to the curve.

The sum of such a series of vectors differs in no way from the sum of a finite number of finite vectors. If $d\mathbf{a}$ represents any one of these small vectors, then by adding them all the resultant $\overrightarrow{AB}$ is obtained. The operation of adding this infinite number of infinitesimal vectors may be represented by an integration sign thus:

$$\overrightarrow{AB} = \int_A^B d\mathbf{a}. \tag{3}$$

If the curve is a closed one, whether a plane curve or not, then A and B coincide and $\overrightarrow{AB} = 0$ or $\int d\mathbf{a}$ around a closed path is zero.

Scalar and Vector Fields and their Addition.

5. Point-Function. Definition of Lamé. If for every position of a point in a region of space a quantity has one or more definite values assigned to it, it is said to be a function of the point, or more concisely, a point-function. We may have both scalar and vector point-functions.

As an **Example of a Scalar Point-Function,** consider the potential at any point due to any distribution of matter M_1 and let its value be V_1. Now consider the potential at the same point P due to any other distribution of matter M_2 and let its value be V_2. Then the potential at P due to both masses together is simply $V_1 + V_2$. This value is found by adding together the two scalars V_1 and V_2.*

* Perhaps the following example of scalar field will be clearer to some minds. Consider a point P and let it be illuminated by a source of light M_1. Evidently every point in the vicinity of the source is illuminated to a greater or lesser extent according to its distance from the source. The illumination or intensity of light at all points of the space considered may be represented by a scalar point-function. Let now another source of light M_2 be brought into the space under consideration. This source produces a certain intensity of illumination at every point of the space, of course including the point P. The total *amount* of illumination now received at the point P is the scalar sum of the amounts it receives from each individual source. This is true of every other point in the field. So that in general in order to find the illumination at any point due to separate sources, one simply adds the

Practical Definition of Continuity of a Scalar Point Function. If, as we go from any point in space to any near adjacent point, the magnitude of the scalar point-function

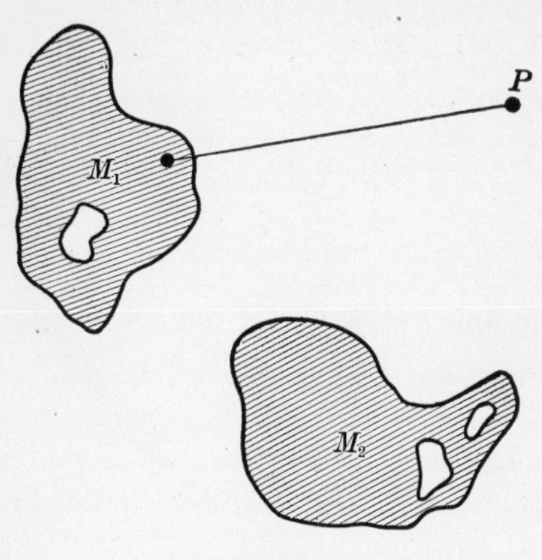

FIG. 5.

suffers no abrupt change, the function is said to be continuous.

As an **Example of a Vector Point-Function** consider the force of attraction at any point P due to the attraction of the mass M_1. This force is evidently a vector quantity, as it has a definite magnitude and a definite direction, so that its representation requires the use of a vector at P; let $\mathbf{F}_1$ be this vector. Similarly let $\mathbf{F}_2$ be the vector representing the force at P due to the matter M_2. Then the force at P due to the combined action of M_1 and M_2 is the vector sum of $\mathbf{F}_1$ and $\mathbf{F}_2$ and must be obtained by the laws of vector addition; *i.e.*, the parallelogram law. If we go from the point P to another point Q in space the magnitudes and direc-

values of the separate intensities at the point due to the separate sources respectively. This constitutes an addition of scalar fields. The fields are here *scalar* fields because we are considering only the *amounts* of the illumination received at any point.

tions of these forces $\mathbf{F}_1$ and $\mathbf{F}_2$ at Q, and hence, in general, their sum, $\mathbf{F}_1 + \mathbf{F}_2$, at Q undergo changes.

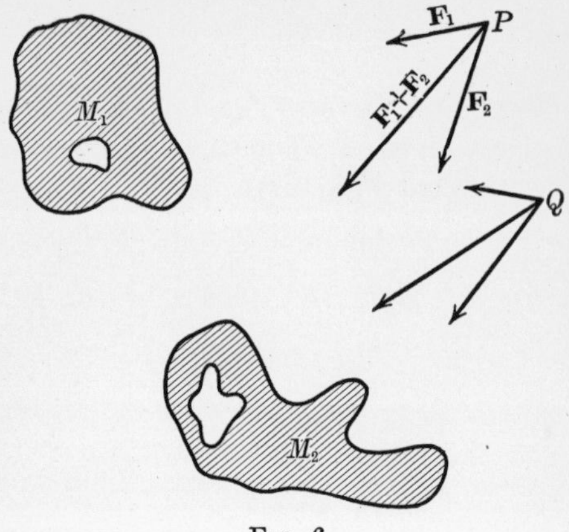

FIG. 6.

Practical Definition of the Continuity of a Vector Point Function. If, as we go from any point in space to any near adjacent point, the direction as well as the magnitude of the vector point-function suffers no abrupt change, the function is said to be continuous.

6. Decomposition of Vectors into Components. From § 4 it is evident that any vector $\mathbf{q}$ may be considered as the sum of any number of component vectors, which when joined end to end, as in vector addition, the first one begins at the origin of $\mathbf{q}$, and the last one ends at the terminus of $\mathbf{q}$. Thus:

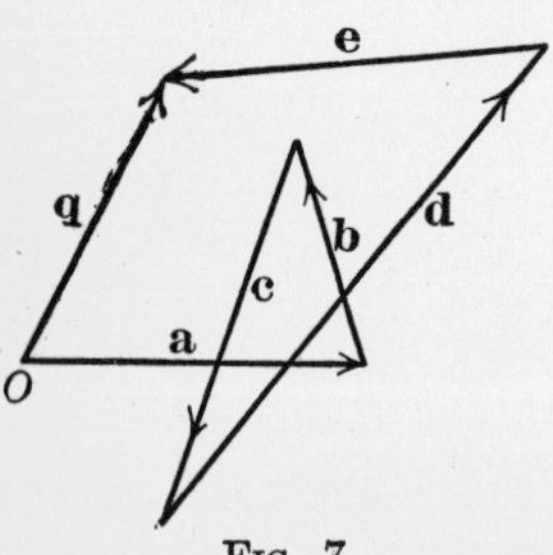

FIG. 7.

$$\mathbf{q} = \mathbf{a} + \mathbf{b} + \mathbf{c} + \mathbf{d} + \mathbf{e}.$$

These vectors need not lie in one plane. Vectors all of which lie in or parallel to the same plane are said to be *coplanar.*

In particular it is often convenient to decompose a vector

into two or three components at right angles to each other; *two* in case all the vectors under consideration are coplanar; *three*, when they are not coplanar.

7. The Three Unit Vectors i j k. Consider the right-handed Cartesian system of axes. The three unit vectors along the $x\ y\ z$ axes are called $\mathbf{i}\ \mathbf{j}\ \mathbf{k}$ respectively. It is evident that any vector $\mathbf{r}$ is equivalent to a certain vector $\overrightarrow{OA}$ along OX, plus a vector $\overrightarrow{AB}$ along OZ, plus a vector $\overrightarrow{BC}$ along OY.

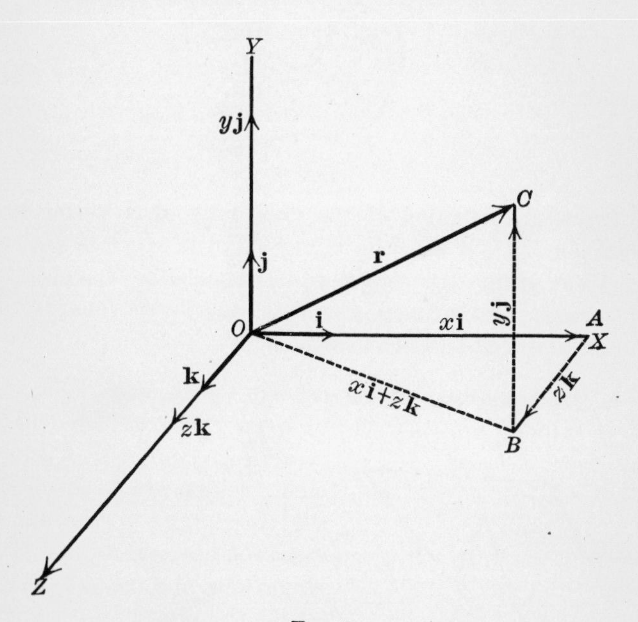

FIG. 8.

In other words, if $x\ y\ z$ denote the magnitudes of these vectors respectively, we may write for any vector $\mathbf{r}$ whose components are x, y, z,

$$\mathbf{r} = x\,\mathbf{i} + y\,\mathbf{j} + z\,\mathbf{k}. \tag{4}$$

$x\mathbf{i}$, $y\mathbf{j}$, and $z\mathbf{k}$ are the three projections of $\mathbf{r}$ along the three axes respectively. If α, β, γ be the direction angles of any vector parallel to OC, then evidently

$$\begin{aligned} x &= r \cos \alpha, \\ y &= r \cos \beta, \\ z &= r \cos \gamma. \end{aligned} \tag{5}$$

This decomposition of a vector into two or three rectangular components is of the utmost importance and is the connecting link between the two or three dimensional Cartesian and Vector Analyses, respectively.

If two vectors are given,

$$\mathbf{a} = a_1\mathbf{i} + a_2\mathbf{j} + a_3\mathbf{k},$$
$$\mathbf{b} = b_1\mathbf{i} + b_2\mathbf{j} + b_3\mathbf{k},$$

their sum is evidently

$$(\mathbf{a} + \mathbf{b}) = (a_1 + b_1)\,\mathbf{i} + (a_2 + b_2)\,\mathbf{j} + (a_3 + b_3)\,\mathbf{k}. \tag{6}$$

This may be extended to any number of vectors and shows that the components of the sum are equal to the sums of the components, so that

$$\sum \mathbf{a} = \mathbf{i} \sum a_1 + \mathbf{j} \sum a_2 + \mathbf{k} \sum a_3. \tag{7}$$

This theorem is of use in the composition of forces. It is possible to resolve any vector $\mathbf{r}$ into three components parallel to *any* three non-coplanar vectors; and such a resolution is easily seen to be unique. Practically, in order to find the rectangular components of a vector, equations (5) are employed, so that

$$\mathbf{r} = r\,(\mathbf{i} \cos \alpha + \mathbf{j} \cos \beta + \mathbf{k} \cos \gamma). \tag{8}$$

If we divide through by the magnitude of $\mathbf{r}$ there remains

$$\frac{\mathbf{r}}{r} = \mathbf{r}_1 = \mathbf{i} \cos \alpha + \mathbf{j} \cos \beta + \mathbf{k} \cos \gamma, \tag{9}$$

so that the rectangular components of a unit vector are always its direction cosines.

By inspection of Fig. 8 it is evident that

$$r^2 = x^2 + y^2 + z^2.$$

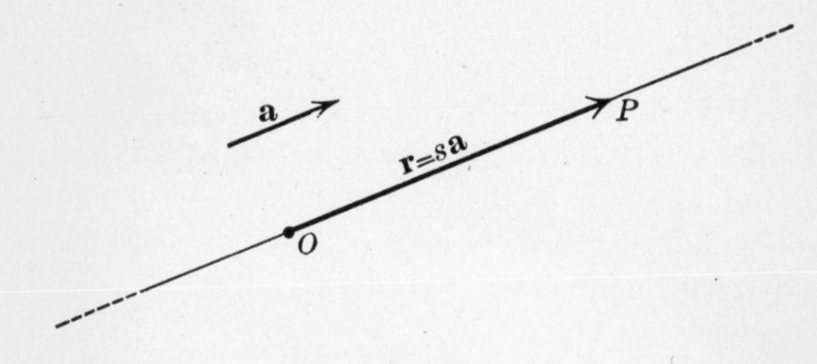

Fig. 9.

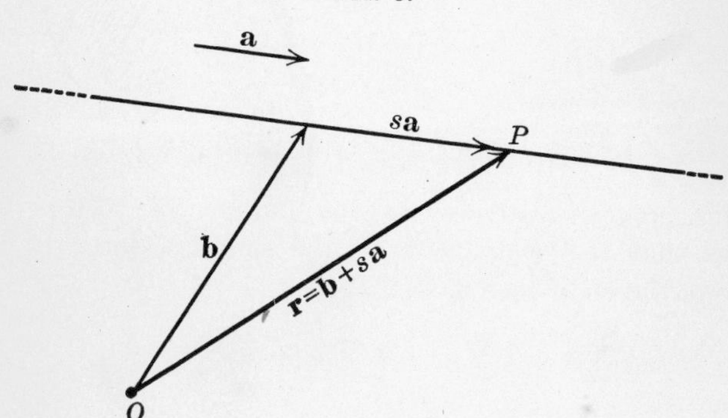

Fig. 10.

Vector Equations.

8. Equations of the Straight Line and Plane. Let **r** be a variable vector, with origin at O, and s a variable scalar; it is then evident on inspection (Fig. 9) that

$$\mathbf{r} = s\,\mathbf{a} \tag{10}$$

is the equation of a straight line passing through the origin and parallel to **a**. It is also easily seen (Fig. 10) that

$$\mathbf{r} = \mathbf{b} + s\,\mathbf{a} \tag{11}$$

is the equation of the straight line through the terminus of **b** and parallel to **a**. By means of equation (11) the equation of a line passing through the ends of any two given vectors **a** and **b** may easily be derived.

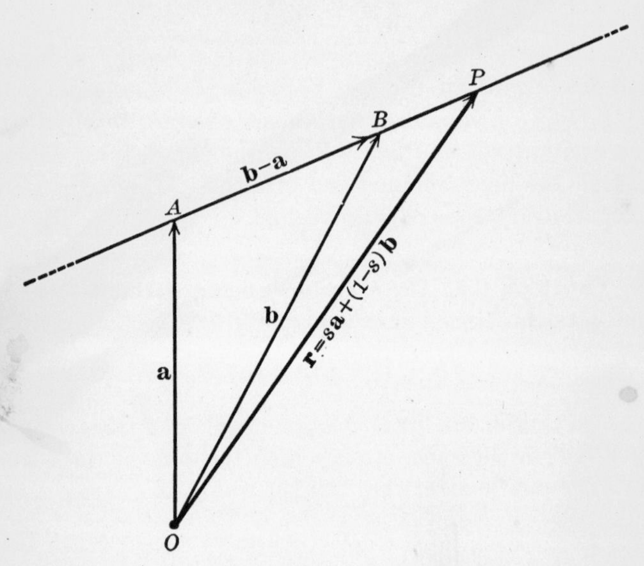

FIG. 11.

The vector AB is $(\mathbf{b} - \mathbf{a})$, hence by equation (11) the line through the terminus of **a** parallel to $(\mathbf{b} - \mathbf{a})$ is

$$\mathbf{r} = \mathbf{a} + t\,(\mathbf{b} - \mathbf{a}),$$

where t is a scalar variable. These equations may be put into the easily remembered forms

$$\mathbf{r} = t\,\mathbf{b} + (1 - t)\,\mathbf{a},$$

and by analogy

$$\mathbf{r} = s\,\mathbf{a} + (1 - s)\,\mathbf{b}. \tag{12}$$

It is evident that if the directions of the coördinate axes be taken along **a** and **b**, then the magnitudes of **a** and **b** respectively are the intercepts the line makes with these axes, the corresponding Cartesian equation being

$$\frac{x}{a} + \frac{y}{b} = 1.$$

All problems in line geometry are now readily solvable. If all the lines of the problem lie in one plane, two, and only two, *arbitrary* non-parallel vectors are chosen and all others expressed in terms of them. For a problem in three dimensions all the lines are expressed in terms of three, and only three, arbitrary non-coplanar vectors.

9. Condition that Three Vectors should Terminate in the Same Straight Line. Putting equation (12) in the form

$$s\,\mathbf{a} + (1 - s)\,\mathbf{b} - \mathbf{r} = 0$$

it is seen that in the linear relation connecting three vectors which end in the same straight line the sum of the coefficients is equal to zero. Or in other words, if

$$x\,\mathbf{a} + y\,\mathbf{b} + z\,\mathbf{c} = 0 \qquad (13)$$

and

$$x + y + z = 0,$$

the three vectors **a**, **b**, and **c** necessarily end in the same straight line, and are said to be *termino-collinear.*

Example. As a simple example of the general method of procedure, let us prove that the diagonals of a parallelogram meet in a point which bisects them both. Take the origin at the corner O, and write down the equation of the diagonals OC and AB in terms of **a** and **b**, the vectors OA and AB. Notice that the origin may be chosen arbitrarily and hence may be taken so as to simplify the equations. Very often, however, it is better not to place the origin at any special or definite point, so that more symmetry is produced in the resulting equations.

The equation of OC is

$$\mathbf{r} = s\,(\mathbf{a} + \mathbf{b}), \qquad \text{by equation (10)}$$

and that of AB is

$$\mathbf{r} = t\,\mathbf{a} + (1 - t)\,\mathbf{b}, \qquad \text{by equation (12)}$$

where s and t are variable scalars. For intersection, both equations must be satisfied by the same value of $\mathbf{r}$; hence, equating,

$$s\,(\mathbf{a} + \mathbf{b}) = t\,\mathbf{a} + (1 - t)\,\mathbf{b}. \qquad (14)$$

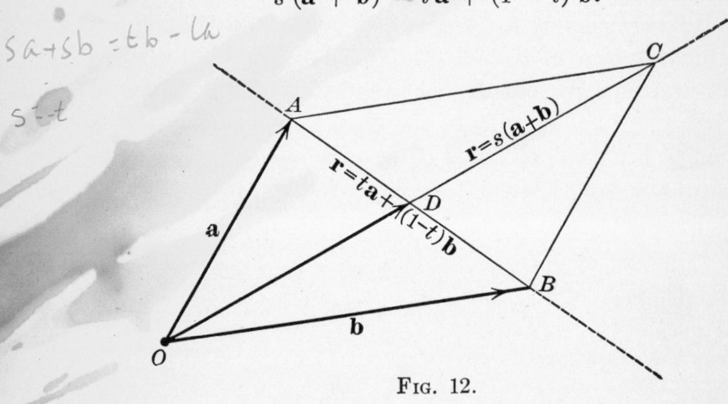

Fig. 12.

This vector equation is actually equivalent to two scalar equations and suffices to determine s and t, for the vector $\mathbf{r}$ to the point of intersection is *uniquely determined* in terms of the vectors $\mathbf{a}$ and $\mathbf{b}$, so that the scalar coefficients of these vectors on both sides of equation (14) must be respectively equal. The coefficients of $\mathbf{a}$ give

$$s = t$$

and of $\mathbf{b}$

$$s = (1 - t).$$

This makes $s = t = \frac{1}{2}$, and the vector to the point of intersection is then, by substituting this value for s in $\mathbf{r} = s\,(\mathbf{a} + \mathbf{b})$,

$$\overrightarrow{OD} = \tfrac{1}{2}\,(\mathbf{a} + \mathbf{b}) = \tfrac{1}{2}\,\overrightarrow{OC}.$$

This *principle* is applicable to any kind of line problem in two or three dimensions. The method of equating the coefficients of the same vector on both sides of an equation is analogous to the conditions for equality of two complex imaginary expressions; that is, if

$$s + it = s' + it',$$

then
$$s = s' \text{ and } t = t'.$$

Example. As an example of the symmetrical method to prove that the medians of a triangle meet in a single point which trisects each of them. Choose any point *not* in the plane of the triangle for origin, and define the triangle by the three vectors **a**, **b**, and **c** from the origin to its vertices A, B, and C. We choose the origin out of the plane of the triangles so that we may use *three* independent vectors instead of but two, as would be necessary if the origin were taken in the same plane.

Then
$$OA' = \tfrac{1}{2}(\mathbf{b} + \mathbf{c}),$$
$$OB' = \tfrac{1}{2}(\mathbf{c} + \mathbf{a}),$$
$$OC' = \tfrac{1}{2}(\mathbf{a} + \mathbf{b}),$$

so that the equation of

AA' is
$$\mathbf{r} = x\,\mathbf{a} + (1 - x)\,\tfrac{1}{2}(\mathbf{b} + \mathbf{c}), \qquad (a)$$

BB' is
$$\mathbf{r} = y\,\mathbf{b} + (1 - y)\,\tfrac{1}{2}(\mathbf{c} + \mathbf{a}), \qquad (b)$$

CC' is
$$\mathbf{r} = z\,\mathbf{c} + (1 - z)\,\tfrac{1}{2}(\mathbf{a} + \mathbf{b}). \qquad (c)$$

Equate the coefficients in (a) and (b) for intersection,

of **a**,
$$x = \tfrac{1}{2}(1 - y),$$
of **b**,
$$\tfrac{1}{2}(1 - x) = y,$$
of **c**,
$$\tfrac{1}{2}(1 - x) = \tfrac{1}{2}(1 - y),$$

so that $x = y = \frac{1}{3}$ and the vector to their point of intersection is

$$\overrightarrow{OD} = \mathbf{r} = \frac{\mathbf{a} + \mathbf{b} + \mathbf{c}}{3},$$

This is evidently the point of intersection of the third line with either of the first two, by symmetry. It is also the

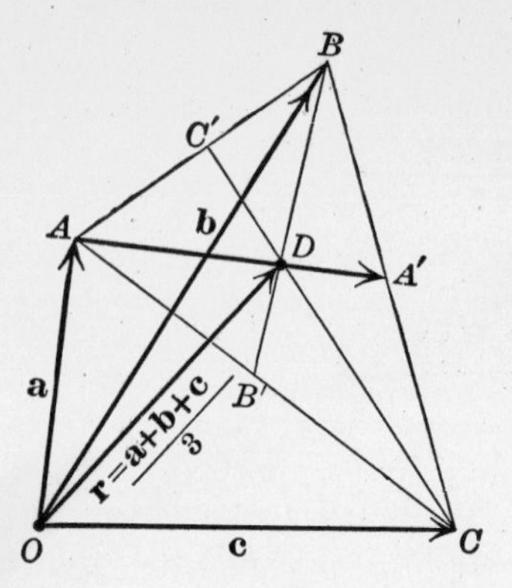

FIG. 13.

mean point of A, B, and C, as explained below. It is the point of trisection, because adding to $\mathbf{a}$, $\frac{2}{3}$ of $\overrightarrow{AA'}$ we obtain the same result, thus:

$$OA' = \tfrac{1}{2}(\mathbf{b} + \mathbf{c}),$$
$$\tfrac{2}{3}AA' = \tfrac{2}{3}[-\mathbf{a} + \tfrac{1}{2}(\mathbf{b} + \mathbf{c})],$$

and $$OD = \mathbf{a} + \tfrac{2}{3}[-\mathbf{a} + \tfrac{1}{2}(\mathbf{b} + \mathbf{c})] = \frac{\mathbf{a} + \mathbf{b} + \mathbf{c}}{3}.$$

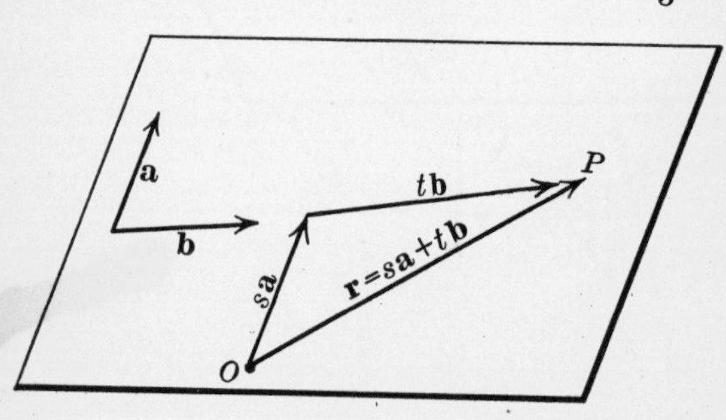

FIG. 14.

By choosing the origin at one of the vertices the symmetry is lost but a gain in directness and shortness is made. In problems involving algebraic coefficients instead of numerical ones the symmetrical method is generally preferable.

10. Equation of a Plane. The vector to any point in the plane determined by the vectors **a** and **b** and passing through the origin is evidently

$$\mathbf{r} = s\,\mathbf{a} + t\,\mathbf{b}, \tag{14}$$

where s and t are two *independent* scalar variables. If the origin be removed to the origin of a vector **c**, through the terminus of which the plane parallel to **a** and **b** passes, then the vector to any point P in the plane is now given by

$$\mathbf{r} = \mathbf{c} + s\,\mathbf{a} + t\,\mathbf{b}. \tag{15}$$

11. To find the equation of a plane passing through the ends of the three non-coplanar vectors a, b, and c, notice that the vectors $(\mathbf{a}-\mathbf{c})$ and $(\mathbf{b}-\mathbf{c})$ evidently lie in the plane. By employing the previous equation (15), the equation may be written

$$\mathbf{r} = \mathbf{c} + s\,(\mathbf{a} - \mathbf{c}) + t\,(\mathbf{b} - \mathbf{c}),$$

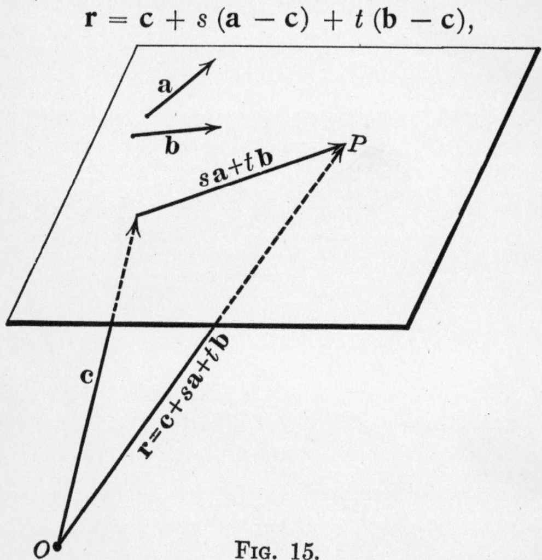

FIG. 15.

which may be put into easily remembered form, analogous to equation (12)

$$\mathbf{r} = s\,\mathbf{a} + t\,\mathbf{b} + (1 - s - t)\,\mathbf{c}. \tag{16}$$

It is evident that if the directions of the coördinate axes be taken along **a**, **b**, and **c**, then the intercepts made by the

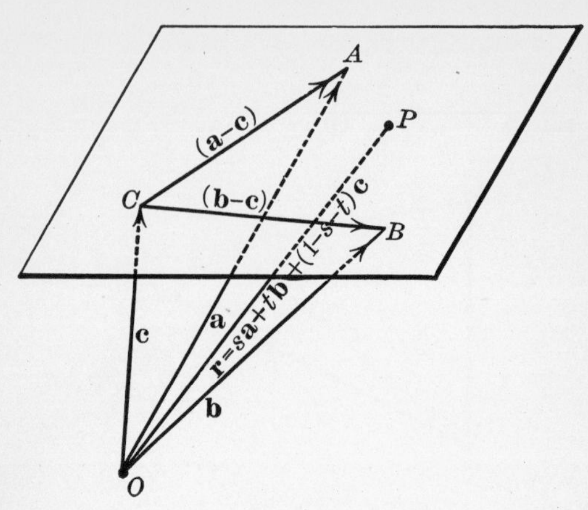

FIG. 16.

plane with these axes are the lengths of **a**, **b**, and **c** respectively, the corresponding Cartesian equation being

$$\frac{x}{a} + \frac{y}{b} + \frac{z}{c} = 1.$$

12. Condition that Four Vectors Terminate in the Same Plane. Rearranging equation (16),

$$s\,\mathbf{a} + t\,\mathbf{b} + (1 - s - t)\,\mathbf{c} - \mathbf{r} = 0,$$

it is seen that whenever there is a linear relation between any four vectors they terminate in one and the same plane if the sum of the coefficients is zero. Or in other words, if

$$x\,\mathbf{a} + y\,\mathbf{b} + z\,\mathbf{c} + w\,\mathbf{d} = 0$$

and

$$x + y + z + w = 0, \tag{17}$$

a, **b**, **c**, and **d** terminate in the same plane and are said to be *termino-coplanar.*

13. To Divide a Line in a Given Ratio. Centroid. To find the value of a vector which divides the distance between two points A and B in a given ratio, m to n say, it is simply necessary to express the vector **r** in the form, evident on inspection,

$$\mathbf{r} = \mathbf{a} + \frac{m}{m+n}(\mathbf{b} - \mathbf{a}) = \frac{n\,\mathbf{a} + m\,\mathbf{b}}{m+n}\,. \tag{18}$$

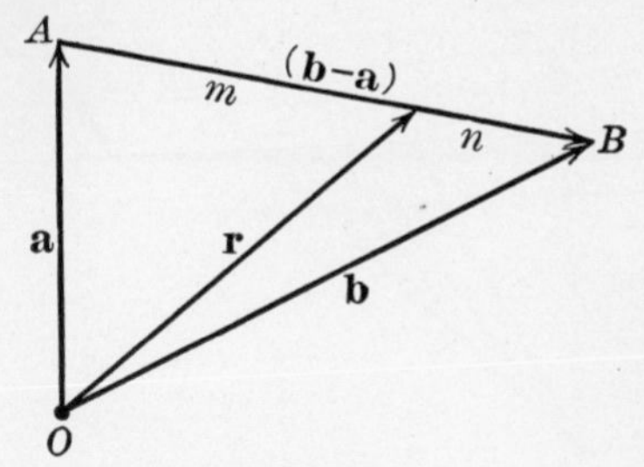

FIG. 17.

It is a well-known result in mechanics that the center of gravity of two masses m_1 and m_2 divides the line joining them inversely as these masses, so that by (18)

$$\mathbf{r}_2 = \frac{m_1\mathbf{a}_1 + m_2\mathbf{a}_2}{m_1 + m_2}$$

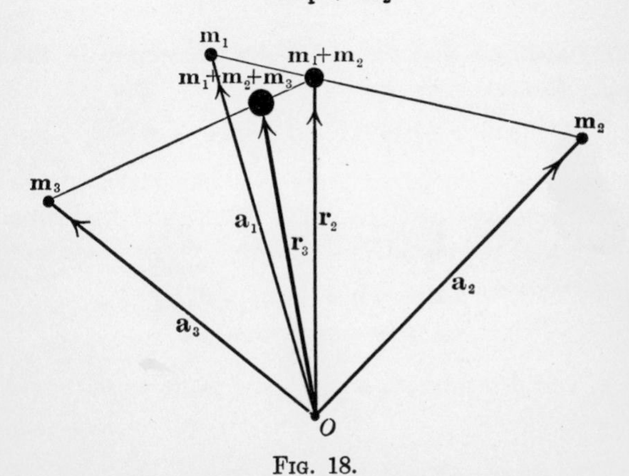

FIG. 18.

is the vector to their center of **mass** or their *centroid*. If now there is a third point $\mathbf{a}_3$ with mass m_3 added to the system, the new centroid will be that of the two masses $\mathbf{r}_2$ with mass $(m_1 + m_2)$ and $\mathbf{a}_3$ with mass m_3, or again by (18),

$$\mathbf{r}_3 = \frac{(m_1 + m_2)\,\mathbf{r}_2 + m_3\mathbf{a}_3}{m_1 + m_2 + m_3} = \frac{m_1\mathbf{a}_1 + m_2\mathbf{a}_2 + m_3\mathbf{a}_3}{m_1 + m_2 + m_3} = \frac{\Sigma\, m\,\mathbf{a}}{\Sigma\, m}. \qquad (19)$$

The generalization is immediate. If $M = \Sigma\, m$ denotes the total mass of the system of particles and $\bar{\mathbf{r}}$ the vector to their center of mass,

$$M\bar{\mathbf{r}} = \Sigma\, m\,\mathbf{a}. \qquad (20)$$

If the masses form a continuous body, the formula becomes

$$\bar{\mathbf{r}} = \frac{\iiint \rho\,\mathbf{a}\,dv}{\iiint \rho\,dv}, \qquad (21)$$

where ρ is the density and dv is the element of volume. The integrations are taken throughout the volume.

If
$$\bar{\mathbf{r}} = \bar{x}\,\mathbf{i} + \bar{y}\,\mathbf{j} + \bar{z}\,\mathbf{k}$$
and
$$\mathbf{a}_n = x_n\mathbf{i} + y_n\mathbf{j} + z_n\mathbf{k},$$

formula (20) breaks up into the three well-known ones for the three coördinates of the center of mass,

$$M\bar{x} = \sum_1^n m x_n,$$
$$M\bar{y} = \sum_1^n m y_n, \qquad (22)$$
$$M\bar{z} = \sum_1^n m z_n.$$

Similarly (21) gives three of the form

$$\bar{x} = \frac{\iiint x\rho\,dx\,dy\,dz}{\iiint \rho\,dx\,dy\,dz}, \text{ etc.} \qquad (23)$$

14. Relations Independent of the Origin. That the center of gravity and therefore all the formulæ just derived are independent of the origin may be shown by the following reasoning.

Taking the origin at O, the vector to R, the center of gravity of the two masses m and n is, by (18),

$$\mathbf{r} = \frac{m\,\mathbf{a} + n\,\mathbf{b}}{m + n}. \tag{24}$$

Now change the origin to O', the new vectors to the masses being $\mathbf{a}'$ and $\mathbf{b}'$ and the vector to the first origin from the new one being $\mathbf{c}$.

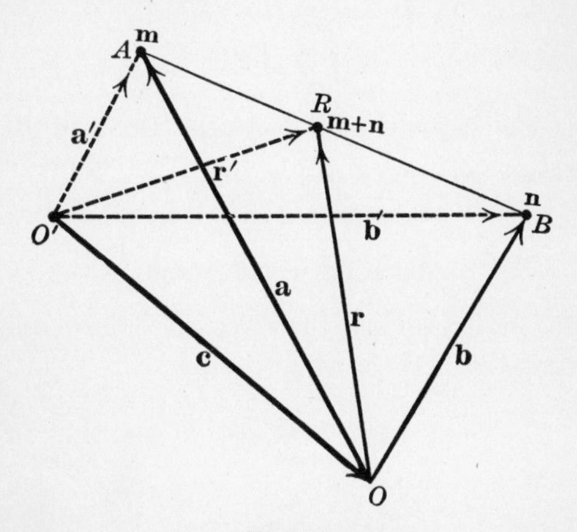

FIG. 19.

The vector to the center of gravity from O' is now given by

$$\mathbf{r}' = \frac{m\,\mathbf{a}' + n\,\mathbf{b}'}{m + n}.$$

But since $\mathbf{a}' = \mathbf{a} + \mathbf{c}$ and $\mathbf{b}' = \mathbf{b} + \mathbf{c}$, this equation may be written

$$\mathbf{r}' = \frac{m\,(\mathbf{c} + \mathbf{a}) + n\,(\mathbf{c} + \mathbf{b})}{m + n} = \frac{m\,\mathbf{a} + n\,\mathbf{b}}{m + n} + \mathbf{c},$$

which says that the new center of gravity is the same as before, as $\mathbf{r}' = \mathbf{r} + \mathbf{c}$.

It will be noticed on writing (24) in the form

$$(m + n)\,\mathbf{r} - m\,\mathbf{a} - n\,\mathbf{b} = 0$$

that the algebraic sum of the scalar coefficients is zero. This leads to the

General Condition for a Relation Independent of the Origin. The necessary and sufficient condition that a linear vector equation represent a relation independent of the origin is that the sum of the scalar coefficients of the equation be equal to zero. Let the equation be

$$m_1\mathbf{a}_1 + m_2\mathbf{a}_2 + \cdots = 0. \tag{25}$$

Change the origin from O to O' by adding a constant vector $\mathbf{l}$, the distance from O to O', to each of the vectors, $\mathbf{a}_1$, $\mathbf{a}_2$, $\mathbf{a}_3$, etc.; the equation then becomes

$$m_1\,(\mathbf{a}_1 + \mathbf{l}) + m_2\,(\mathbf{a}_2 + \mathbf{l}) + \cdots = 0$$

or $$m_1\mathbf{a}_1 + m_2\mathbf{a}_2 + \cdots + \mathbf{l}\,(m_1 + m_2 + \cdots) = 0.$$

If this is to be independent of the origin, *i.e.*, the same as (25), the coefficient of $\mathbf{l}$ must vanish, or

$$m_1 + m_2 + \cdots = 0.$$

EXERCISES AND PROBLEMS.

1. Prove that the vectors

$$\pm\,\mathbf{a} \pm \mathbf{b} \pm \mathbf{c}$$

when drawn from a common origin terminate at the vertices of a parallelopiped.

2. A person traveling eastward at a rate of 3 miles an hour finds that the wind seems to blow directly from the north; on doubling his speed it appears to come from the northeast. Find the vector wind velocity.

3. A ship whose head is pointing due south is steaming across a current running due west; at the end of two hours it is found that the ship has gone 36 miles in the direction 15° west of south. Find the velocities of the ship and current, graphically and analytically.

4. A weight W hangs by a string and is pushed aside by a horizontal force until the string makes an angle of 45° with the vertical. Find the horizontal force and the tension of the string.

5. A vector $\mathbf{r}$ is the resultant of two vectors $\mathbf{a}$ and $\mathbf{b}$ which make angles of 30° and 45° with it on opposite sides. How large are the latter vectors?

6. A car is running at 14 miles an hour and a man jumps from it with a velocity of 8 feet per second in a direction making an angle of 30° with the direction of the car's motion. What is his velocity relative to the ground?

7. Verify, by *drawing*, the truth of the laws of association and commutation, taking a number of vectors, $\mathbf{a}$, $\mathbf{b}$, $\mathbf{c}$, $\mathbf{d}$, etc., to scale, and show that the resultant is independent of the order of addition or subtraction.

8. Given the vector

$$\mathbf{r} = a_1\mathbf{i} + a_2\mathbf{j}$$

derive the vector of same length perpendicular to it through the origin.

Derive the vector perpendicular to the one you find. Compare with the original one.

9. Find the relative motion of two particles moving with the same speed v, one of which describes a circle of radius a while the other moves along a diameter.

10. Two particles move with speeds v and $2v$ respectively in opposite directions, in the circumference of a circle. In what positions is their relative velocity greatest and least, and what values has it at those positions?

11. Draw the vectors

$$\begin{aligned}
\mathbf{a} &= 6\,\mathbf{i} - 4\,\mathbf{j} + 10\,\mathbf{k} \\
\mathbf{b} &= -6\,\mathbf{i} + 4\,\mathbf{j} - 10\,\mathbf{k} \\
\mathbf{c} &= 4\,\mathbf{i} - 6\,\mathbf{j} - 10\,\mathbf{k} \\
\mathbf{d} &= \phantom{-6\,\mathbf{i} -{}} 10\,\mathbf{j} + 4\,\mathbf{k}
\end{aligned}$$

Find their sum graphically and analytically.

12. The equation

$$(\mathbf{r} - \mathbf{a})_0 = (\mathbf{r} - \mathbf{b})_0$$

represents the plane bisecting at right angles the line AB.

13. Find the equation of the locus of a point equidistant from two fixed planes.

14. The line which joins one vertex of a parallelogram to the middle point of an opposite side trisects the diagonal.

15. To find a line which passes through a given point and cuts two given lines in space.

16. If

$$x\,\mathbf{a} + y\,\mathbf{b} = 0$$

and

$$x + y = 0$$

show that **a** and **b** are equal in magnitude and direction. Or what is the same thing, that measured from the same origin, **a** and **b** end at the same point.

17. If

$$x\,\mathbf{a} + y\,\mathbf{b} + z\,\mathbf{c} = 0$$

and

$$x + y + z = 0$$

show that **a**, **b**, and **c** terminate in the same straight line; they are then said to be termino-collinear.

18. If

$$x\,\mathbf{a} + y\,\mathbf{b} + z\,\mathbf{c} + w\,\mathbf{d} = 0$$

and

$$x + y + z + w = 0$$

show that **a**, **b**, **c**, and **d** terminate in the same plane; they are then said to be termino-coplanar.

19. A triangle may be constructed whose sides are equal and parallel to the medians of any given triangle.

20. Given a quadrilateral in space. Find the middle point of the line which joins the middle points of the diagonals. Find the middle point of the line joining the middle points of two opposite sides. Show that these two points are the same and coincide with the center of gravity of a system of equal masses placed at the vertices of the quadrilateral.

21. Discuss the conditions imposed upon three, four, or five vectors if they satisfy *two* equations, the sum of the coefficients in each of which is zero.

22. Take a number of points at random on a sheet of paper, assigning arbitrary masses to them. Verify by drawing that their center of mass is independent of the origin chosen in finding it.

23. If a system of masses, each mass concentrated at a point, be divided into a number of partial systems, and each of these be replaced by its resultant mass, then the new system has the same center of mass as the original one.

24. A cardboard square is bent along a diagonal until the two parts are at right angles. Find the position of the center of gravity.

25. Forces acting at a point O are represented by OA, OB, OC, ..., ON. Show that if they are in equilibrium O is the centroid of the points A, B, C, ..., N.

26. The middle points of the lines which join the points of bisection of the opposite sides of a quadrilateral coincide whether the four sides be in the same plane or not.

27. The bisectors of the angles of a triangle meet in a point which trisects each of them.

Employ unit vectors along two of the sides as independent vectors. The bisectors are then $\mathbf{a}_1 + \mathbf{b}_1$, etc.

28. If two forces acting at a point O are represented by the vectors $n\,\mathbf{a}$ and $\mathbf{b}$ their resultant is represented in magnitude and direction by the vector $(n + 1)\,\overrightarrow{OG}$, the point G being taken on AB so that $BG = nAG$.

This allows the resultant of two forces to be drawn knowing one and *part* of another.

29. If two forces are equal to $n.OA$ and $m.OB$, the resultant passes through the point G determined so that $\frac{BG}{AG} = \frac{n}{m}$ and is equal to $(m + n)\,\overrightarrow{OG}$ in magnitude.

30. Forces $\mathbf{F}_1$, $\mathbf{F}_2$, ..., $\mathbf{F}_n$, acting in a plane at O are in equilibrium. Any transversal cuts their lines of action in points L_1, L_2, ..., L_n; and a length $\overrightarrow{OL_i}$ is positive when in the same direction as $\overrightarrow{OF_i}$. Prove that

$$\sum \frac{F}{OL} = 0.$$

31. Show that the resultant of any number of concurrent forces, $\mathbf{F}_1$, $\mathbf{F}_2$, $\mathbf{F}_3$, ... may be found thus: measure off *any* lengths l_1, l_2, l_3, ... from the point of meeting along them respectively; place at the ends of these lines particles of masses proportional to $\frac{F_1}{l_1}$, $\frac{F_2}{l_2}$,

$\frac{F_3}{l_3}$, . . . ; let G be the center of gravity of these particles; then OG is the line of action of the resultant of the given forces and its magnitude is

$$OG \times \sum \frac{F}{l}.$$

32. A particle placed at O is acted upon by forces represented in magnitudes and directions by the lines $OA_1, OA_2, \ldots OA_n$, which join O to any fixed points $A_1, A_2, \ldots A_n$; where must O be placed so that the magnitude of the resultant force may be constant?

Ans. If r represent the magnitude of the resultant, O may be placed anywhere on a sphere of radius $\frac{r}{n}$ described around the centroid of the fixed points as center.

33. $ABCD$ is a quadrilateral of which A and C are opposite vertices. Two forces acting at A are represented by the sides AB and AD; two at C by CB and CD. Prove that the resultant is represented in magnitude and direction by four times the line joining the middle points of the diagonals of the quadrilateral.

34. Show that the resultant of the three vector diagonals of a parallelopiped meeting at a point O is represented by twice the diagonal of the parallelopiped drawn from the same point.

35. If through any point within a parallelogram, parallels be drawn to the sides, the corresponding diagonals of the two new parallelograms thus formed and of the original one meet in a point.

36. The middle points P, Q, R of the diagonals of any complete quadrilateral $ABCDEF$ are collinear.

37. Any point O is joined to the vertices of a parallelogram; show that the sum of the vectors to the vertices is four times the vector to the intersection of the diagonals.

What conclusion do you derive from this fact?

38. $ABCDEFA$ is a regular hexagon. Show that the resultant of the forces represented by AB, $2\,AC$, $3\,AD$, $4\,AE$, $5\,AF$ is represented by a vector of magnitude $\sqrt{351}\,AB$, and find its direction.

39. $ABCDEFA$ is a regular hexagon. Find the resultant of the forces represented by the lines AB, AC, AD, AE, AF.

40. O is any point in the plane of a triangle ABC, and D, E, F are the middle points of the sides. Show that the system of forces OA, OB, OC is equivalent to the system OD, OE, OF.

41. ABC is a triangle with a right angle at A; AD is the perpendicular on BC. Prove that the resultant of forces $\frac{1}{AB}$ acting along AB and $\frac{1}{AC}$ acting along AC is $\frac{1}{AD}$ acting along AD.

42. $P_1, P_2, \ldots P_n$ are points which divide the circumference of a circle into n equal parts. If a particle G lying on the circumference be acted upon by forces represented by $GP_1, GP_2, \ldots GP_n$, show that the magnitude of the resultant is constant wherever G is taken on that circumference.

It is $n \times \overline{OG}$, O being the center of the circle.

43. If O be the center of the circumscribed circle of a triangle ABC, and L the intersection of the perpendiculars from the vertices on the sides, prove that the resultant of forces represented by LA, LB, LC will be represented in magnitude and direction by $2\,LO$.

44. D is a point in the plane of the triangle ABC, and I is the center of its inscribed circle. Show that the resultant of the vectors $a\overrightarrow{AD}, b\overrightarrow{BD}, c\overrightarrow{CD}$ is $(a + b + c)\,\overrightarrow{ID}$, where a, b, c are the lengths of the sides of the triangle.

45. The chords APB and CPD of a circle intersect at right angles. Show that the resultant of $\overline{PA}, \overline{PB}, \overline{PC}$, and $\overline{PD}$ is represented by twice the vector $\overline{PO}$, where O is the center of the circle.

46. Prove that the mean center of a tetrahedron is (*a*) the intersection of bisectors of opposite edges; (*b*) the intersection of lines joining the vertices to the mean points of the opposite faces. Show that the former lines bisect one another, and that the latter quadrisect one another.

47. A, B, and C being three given points in a plane show that any point in this plane can be made their centroid by giving suitable weights to these points.

48. Show that the medians of a triangle intersect in a point which is the mean center of the vertices A, B, C with weights 1, 1, 1; that the altitudes intersect in a point which is the centroid of the vertices with weights, $\tan A$, $\tan B$, $\tan C$, respectively; that the bisectors intersect in a point which is the centroid of the vertices with weights equal to the lengths of the opposite sides.

CHAPTER II.

SCALAR AND VECTOR PRODUCTS OF TWO VECTORS.

The Scalar or Dot Product.

15. The Scalar Product of two vectors **a** and **b**, denoted by **a·b**, *S***ab**, **ab** or (**ab**) by various writers, is a scalar defined by the equation

$$\mathbf{a}\cdot\mathbf{b} = a\,b\cos(\mathbf{ab}) = \mathbf{b}\cdot\mathbf{a}. \qquad (26)$$

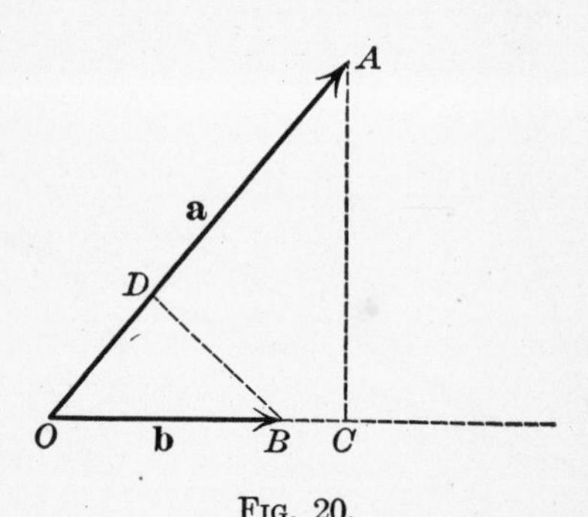

Fig. 20.

This equation shows that the scalar product may be looked upon as the product of the length of one of the two vectors multiplied by the projection of the other upon it, or

$$\overline{OA} \times \overline{OD} = \overline{OB} \times \overline{OC}.$$

Evidently, if the two vectors **a** and **b** are perpendicular to each other $\cos(\mathbf{ab}) = 0$ and their scalar product is zero. The condition, then, of perpendicularity of two finite vectors is that their scalar product be zero.

Or, if $\qquad \mathbf{a}\cdot\mathbf{b} = 0$, then $\mathbf{a} \perp \mathbf{b}$. $\qquad (27)$

If **a** and **b** are parallel vectors, cos (**ab**) = 1 and

$$\mathbf{a}\cdot\mathbf{b} = a\,b,$$

and in particular if $b = a$,

$$\mathbf{a}\cdot\mathbf{a} = a^2.$$

The scalar product of a vector into itself is often written as the square of the vector, thus,

$$\mathbf{a}\cdot\mathbf{a} = \mathbf{a}^2.$$

In general, to obtain the magnitude of a vectorial expression it is only necessary to square it, and the result is the square of its absolute value or magnitude.

The Scalar Product Obeys the Ordinary Laws of Multiplication. Consider the two vectors **c** and **d** as well as their sum (**c** + **d**). Consider also their projections upon any other vector **b**.

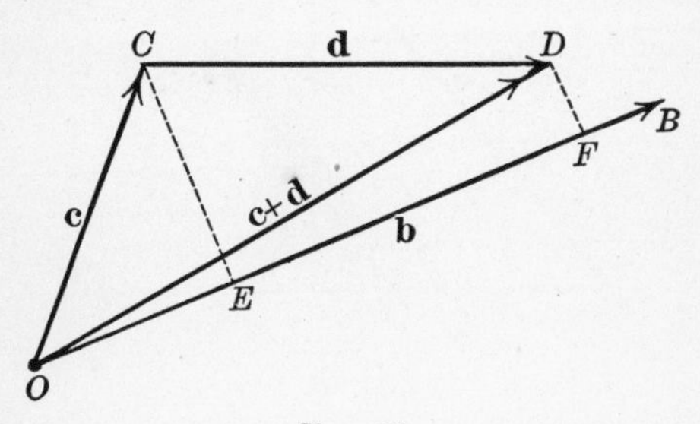

FIG. 21.

The projection of **c** on **b** is OE, the projection of **d** on **b** is EF, the projection of (**c** + **d**) on **b** is OF; hence

$$\mathbf{c}\cdot\mathbf{b} + \mathbf{d}\cdot\mathbf{b} = (\mathbf{c} + \mathbf{d})\cdot\mathbf{b} = \mathbf{b}\cdot(\mathbf{c} + \mathbf{d}). \qquad (28)$$

This result is easily extended to the scalar product of the sums of any number of vectors.

The application of these results to the unit vectors **i**, **j**, and **k** is of great importance, giving immediately

$$\mathbf{i}\cdot\mathbf{i} = \mathbf{j}\cdot\mathbf{j} = \mathbf{k}\cdot\mathbf{k} = \mathbf{i}^2 = \mathbf{j}^2 = \mathbf{k}^2 = 1,$$
$$\mathbf{i}\cdot\mathbf{j} = \mathbf{j}\cdot\mathbf{i} = \mathbf{j}\cdot\mathbf{k} = \mathbf{k}\cdot\mathbf{j} = \mathbf{k}\cdot\mathbf{i} = \mathbf{i}\cdot\mathbf{k} = 0. \qquad (29)$$

If the two vectors **a** and **b** be given in terms of their coördinates,

$$\mathbf{a} = a_1\mathbf{i} + a_2\mathbf{j} + a_3\mathbf{k}$$

and

$$\mathbf{b} = b_1\mathbf{i} + b_2\mathbf{j} + b_3\mathbf{k},$$

then, by (28) and (29),

$$\begin{aligned}\mathbf{a}\cdot\mathbf{b} &= (a_1\mathbf{i} + a_2\mathbf{j} + a_3\mathbf{k})\cdot(b_1\mathbf{i} + b_2\mathbf{j} + b_3\mathbf{k}) \\ &= a_1b_1 + a_2b_2 + a_3b_3. \qquad (30)\end{aligned}$$

If $\mathbf{a}_1$ and $\mathbf{b}_1$ are unit vectors, their projections on the three axes are equal to their direction cosines; and since in this case $\mathbf{a}_1\cdot\mathbf{b}_1 = \cos(\mathbf{a}_1\mathbf{b}_1)$, then, by (30),

$$\begin{aligned}\mathbf{a}_1\cdot\mathbf{b}_1 = \cos(\mathbf{a}_1\mathbf{b}_1) &= \cos(\mathbf{a}_1\mathbf{i})\cos(\mathbf{b}_1\mathbf{i}) + \cos(\mathbf{a}_1\mathbf{j})\cos(\mathbf{b}_1\mathbf{j}) \\ &+ \cos(\mathbf{a}_1\mathbf{k})\cos(\mathbf{b}_1\mathbf{k}),\end{aligned}$$

the familiar formula of Cartesian geometry for the angle between two lines in terms of their direction cosines.

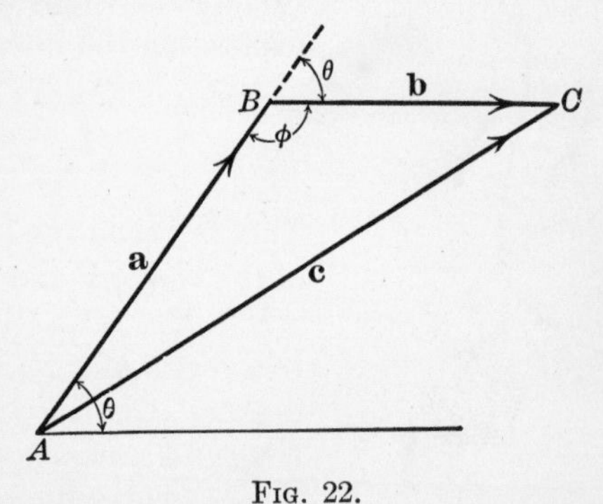

FIG. 22.

The well-known and useful formula giving directly the magnitude of the resultant of any two vectors in terms of their magnitudes and the angle between them, may be derived in the following manner. In the triangle ABC

$$\mathbf{c} = \mathbf{a} + \mathbf{b}.$$

Squaring to find its magnitude,

$$\mathbf{c}\cdot\mathbf{c} = \mathbf{c}^2 = c^2 = (\mathbf{a} + \mathbf{b})\cdot(\mathbf{a} + \mathbf{b}) = \mathbf{a}\cdot\mathbf{a} + 2\,\mathbf{a}\cdot\mathbf{b} + \mathbf{b}\cdot\mathbf{b}$$

or
$$c^2 = a^2 + 2\,a\,b\cos(\mathbf{ab} \text{ or } \theta) + b^2$$

and
$$c^2 = a^2 - 2\,a\,b\cos(\phi) + b^2,$$

where ϕ is the supplement to the angle between **a** and **b**.

16. Line-Integral of a Vector. The scalar product plays a very important rôle in mechanics and physics. For example, the work done by a force **F** in the displacement $d\mathbf{r}$ is by definition

$$F\,dr\cos(\mathbf{F}\,d\mathbf{r}) = \mathbf{F}\cdot d\mathbf{r}.$$

If the force is known in direction and magnitude for every point of its path, the work done in overcoming the forces from A to B may be found by evaluating the integral

$$W = \int_A^B \mathbf{F}\cdot d\mathbf{r}. \qquad (31)$$

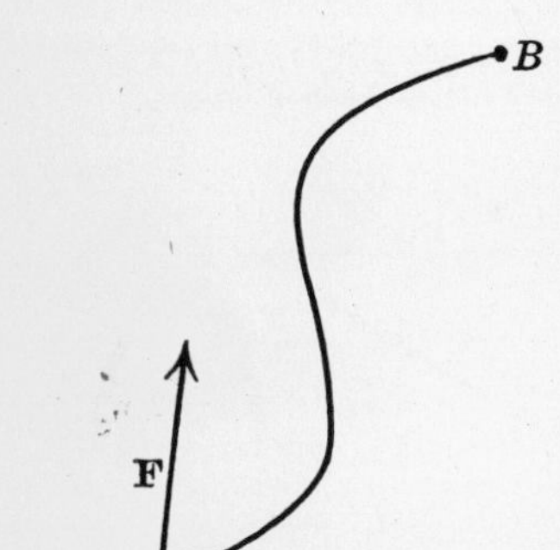

Fig. 23.

This is called the line-integral of the vector **F** along the curve AB. The term "line-integral of a vector along a curve" thus denotes the integral of the *tangential component* along it, unless expressly stated otherwise.

If **q** denote the vector velocity at any point of a fluid, the integral

$$C = \int \mathbf{q}\cdot d\mathbf{r}$$

over any path in the fluid is called the *circulation* along that path. If $\mathbf{e}$ denote the electric force at any point in space, the integral

$$E = \int \mathbf{e} \cdot d\mathbf{r}$$

taken along any path gives the *electro-motive* force along that path. This kind of an integral is thus of great importance in all branches of physics.

17. Surface-Integral of a Vector. As another example, imagine a surface S drawn in any vector field; for example, in a moving fluid. Let $\mathbf{q}$ be the vector velocity, determinate

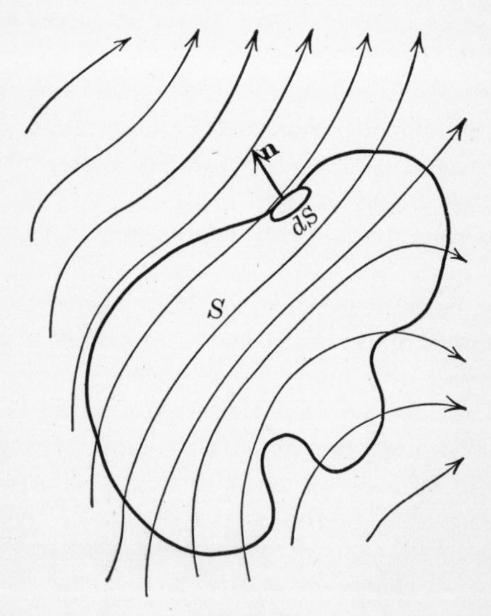

FIG. 24.

at every point in the region considered. The lines of flow of the fluid are therefore known and may be drawn. The amount of liquid which passes outward through the element dS in unit time at any point on the surface is the outward normal component of $\mathbf{q}$ multiplied by the area dS, or

$$q \cos(\mathbf{nq})\, dS = \mathbf{q}\cdot\mathbf{n}\, dS$$

where $\mathbf{n}$ is the unit *outward* drawn normal to dS. The total outward flux through the surface is, then, the surface integral.

$$\text{Total Flux} = \int\!\!\int_S \mathbf{q}\cdot\mathbf{n}\, dS \tag{32}$$

taken over the surface in question. It may easily be seen that in this example the vector $\mathbf{q}$ may be any physical vector such as electric force, magnetic force, gravitational force, or flux of heat, and others.

The term surface-integral of a vector over any surface will in the following denote the integral of the outward normal component over the surface, unless otherwise expressly stated in the context.

The surface integral (32) expresses a very simple fact. If, for instance, we know the motion of every part of a fluid, it should be possible, at least theoretically, to find out how much of the fluid leaves or enters a given region by considering how much passes through every part of the bounding surface of the region and adding the results together. To find the amount passing through any element of the surface we must evidently consider only the normal component of the current of fluid. The tangential component of the current does *not* pass *through* the surface. The integral is the mathematical expression of this conception and represents the total outward flux through the surface S. Of course if the flow is inwards the result will be negative, and if as much flows outwards through one portion of the surface as there flows inwards elsewhere the result will be zero.

The Vector or Cross Product.

18. The vector product of two vectors **a** and **b** is a *vector*, written $\mathbf{a}\times\mathbf{b}$ (in distinction from $\mathbf{a}\cdot\mathbf{b}$, the dot product), also $V\,\mathbf{ab}$ or $[\mathbf{ab}]$ by different authors, and is defined by the equation

$$\mathbf{a}\times\mathbf{b} = \boldsymbol{\epsilon}\, a\, b \sin(\mathbf{ab}) = -\,\mathbf{b}\times\mathbf{a}, \tag{33}$$

where $\boldsymbol{\epsilon}$ is a vector, normal to the plane of **a** and **b** and so directed that as you turn the first named vector **a** into the second one **b**, $\boldsymbol{\epsilon}$ points in the direction that a right-handed screw (cork-screw) would progress if turned in this same

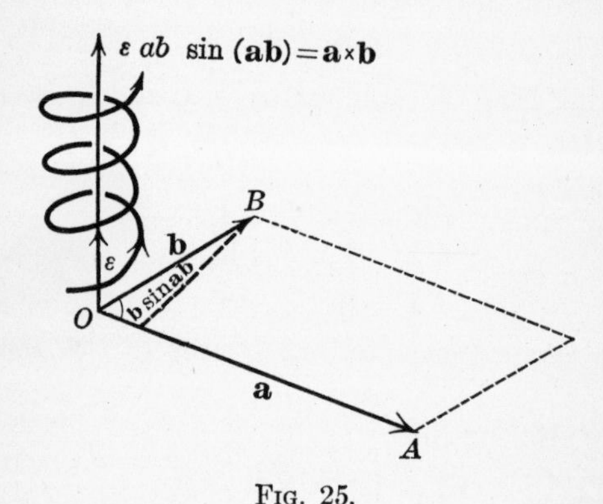

FIG. 25.

manner. In other words, $\mathbf{a}\times\mathbf{b}$ is a vector perpendicular to both **a** and **b** and whose magnitude may be represented by the *area of the parallelogram of which* **a** *and* **b** *are the adjacent sides.* The sense of this vector is purely conventional but is taken to conform with the more usual system of axes, *i.e.*, the right-handed one.

According to this convention if the factor **b** came first instead of **a**, in the product, the only difference would be in the reversal of the sense of $\boldsymbol{\epsilon}$, so that

$$\mathbf{a}\times\mathbf{b} = -\,\mathbf{b}\times\mathbf{a}.$$

It is in this change of sign, when the order of the factors is changed, that the vector product differs from the product of ordinary algebraic or scalar quantities. It is therefore necessary when manipulating vector products to preserve the order of the factors unchanged or, at every change of order, to introduce a minus sign as a factor.

In particular if **a** and **b** be finite vectors and

$$\mathbf{a}\times\mathbf{b} = 0, \quad \text{then} \quad \mathbf{a} \parallel \mathbf{b}$$

as the sine of their included angle must be zero. This, then, is the condition for parallelism of the two vectors **a** and **b**. Since any vector is parallel to itself,

$$\mathbf{a}\times\mathbf{a} \equiv 0. \tag{34}$$

Remembering that the unit vectors **i**, **j**, and **k** are mutually perpendicular, it follows immediately from the definition that

$$\begin{aligned} \mathbf{j}\times\mathbf{k} &= \mathbf{i} = -\,\mathbf{k}\times\mathbf{j}, \\ \mathbf{k}\times\mathbf{i} &= \mathbf{j} = -\,\mathbf{i}\times\mathbf{k}, \\ \mathbf{i}\times\mathbf{j} &= \mathbf{k} = -\,\mathbf{j}\times\mathbf{i}. \end{aligned} \tag{35}$$

Notice the cyclical order of the factors in the above equations.

We have also, by (34),

$$\mathbf{i}\times\mathbf{i} = \mathbf{j}\times\mathbf{j} = \mathbf{k}\times\mathbf{k} \equiv 0.$$

19. Distributive Law for Vector Products. It is obvious from the definition of $\mathbf{a}\times\mathbf{b}$ that

$$\mathbf{a}\times\mathbf{b} = \mathbf{a}'\times\mathbf{b}, \tag{36}$$

where $\mathbf{a}'$ is the component of **a** $\perp$ to **b**. Because in Fig. (26), as $\mathbf{a}'$ and **b** are in the same plane as **a** and **b**, $\boldsymbol{\epsilon}$ is the same as before, and $\mathbf{a}_0' = \mathbf{a}_0 \sin\phi$. We may also say that the vector product of **b** with the component of **a** parallel to **b** is zero. So that in any vector product we may, if we wish, replace one of the vectors by its normal component to the other, and *vice versa*, without changing the value of the product.

Keeping this in mind, we may prove that the distributive law holds for vector products, or, in symbols, that

$$(\mathbf{a} + \mathbf{b})\times\mathbf{c} = \mathbf{a}\times\mathbf{c} + \mathbf{b}\times\mathbf{c}. \qquad (37)$$

where **a** and **b** are any two vectors. Let **c** be drawn (Fig. 27) ⊥ to the plane of the paper at O and towards the reader. Let $\mathbf{a}'$ and $\mathbf{b}'$ be the components of **a** and **b** ⊥ to **c** and hence lying in the plane of the paper. The vectors

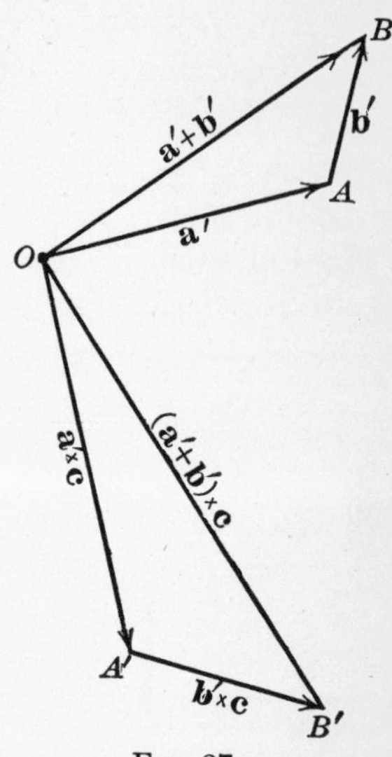

Fig. 26.

Fig. 27.

$\mathbf{a}'\times\mathbf{c}$ and $\mathbf{b}'\times\mathbf{c}$ will also lie in the plane of the paper perpendicular to $\mathbf{a}'$ and $\mathbf{b}'$ respectively.

$$\text{Since} \quad \frac{A'B'}{OA'} = \frac{(\mathbf{b}'\times\mathbf{c})_0}{(\mathbf{a}'\times\mathbf{c})_0}{}^{*} = \frac{b'c\sin\frac{\pi}{2}}{a'c\sin\frac{\pi}{2}} = \frac{b'}{a'} = m,$$

the triangles OAB and $OA'B'$ are similar, hence $\dfrac{OB}{OB'} = m$ and OB' is ⊥ to OB. Consequently

$$OB' = (\mathbf{a}' + \mathbf{b}')\times\mathbf{c} = \overrightarrow{OA'} + \overrightarrow{A'B'} = \mathbf{a}'\times\mathbf{c} + \mathbf{b}'\times\mathbf{c}.$$

We may now replace $\mathbf{a}'$ and $\mathbf{b}'$ by **a** and **b** according to (36) above, so that

$$(\mathbf{a} + \mathbf{b})\times\mathbf{c} = \mathbf{a}\times\mathbf{c} + \mathbf{b}\times\mathbf{c}.$$

* See equation (1) for notation.

If **c** itself be considered to be made up of two vectors **e** and **f**, then by the same reasoning

$$\mathbf{a}\times(\mathbf{e}+\mathbf{f}) = \mathbf{a}\times\mathbf{e} + \mathbf{a}\times\mathbf{f}$$

and

$$\mathbf{b}\times(\mathbf{e}+\mathbf{f}) = \mathbf{b}\times\mathbf{e} + \mathbf{b}\times\mathbf{f},$$

so that

$$(\mathbf{a}+\mathbf{b})\times(\mathbf{e}+\mathbf{f}) = \mathbf{a}\times\mathbf{e} + \mathbf{a}\times\mathbf{f} + \mathbf{b}\times\mathbf{e} + \mathbf{b}\times\mathbf{f} \tag{38}$$

and so on for any number of vectors.

Physical Proof of the Distributive Law. — It is interesting to prove the distributive law for vector products by means of the following hydrostatic theorem. It is well

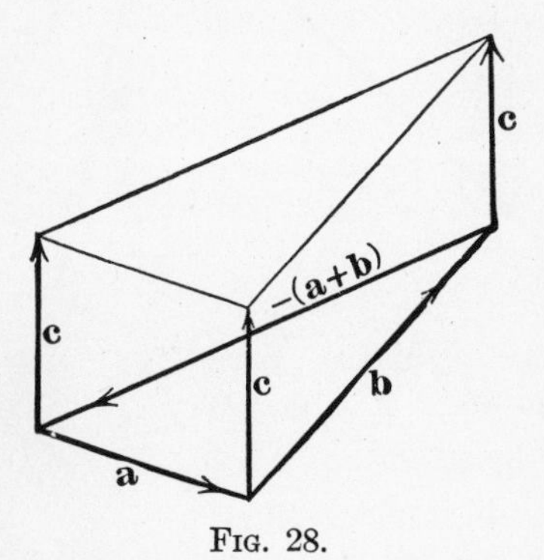

FIG. 28.

known that any closed polyhedral surface immersed in a fluid is in equilibrium under the normal hydrostatic pressures exerted upon its faces by the liquid. These pressures produce forces normal to the faces of the polyhedron which are proportional to their areas and may therefore be represented by vectors perpendicular to them, the length of each one being proportional to the total pressure on the face to which it is perpendicular. The condition for equilibrium is then that the sum of these vectors be zero, *i.e.*, that they have no resultant. This result is seen to be also true for any *curved* surface by considering it as the limiting case of a polyhedron with an infinite number of infinitely small, plane facets. Let **a**, **b** and $-(\mathbf{a}+\mathbf{b})$ be the three sides of a triangle, taken in order. Form the prism of which this triangle and any third vector **c** is the slant height or edge. The areas of the lateral faces of this prism are respectively, viewing them from the outside,

$$\mathbf{a}{\times}\mathbf{c} \quad \mathbf{b}{\times}\mathbf{c} \quad \text{and} \quad -(\mathbf{a}+\mathbf{b}){\times}\mathbf{c};$$

the areas of the end faces are similarly

$$\tfrac{1}{2}\,\mathbf{a}{\times}\mathbf{b} \quad \text{and} \quad -\tfrac{1}{2}\,\mathbf{a}{\times}\mathbf{b}.$$

Now by the preceding hydrostatic theorem the vector sum of the faces of any closed surface is zero, hence

$$\mathbf{a}{\times}\mathbf{c} + \mathbf{b}{\times}\mathbf{c} - (\mathbf{a}+\mathbf{b}){\times}\mathbf{c} + \tfrac{1}{2}\,\mathbf{a}{\times}\mathbf{b} - \tfrac{1}{2}\,\mathbf{a}{\times}\mathbf{b} = 0,$$

giving again $$(\mathbf{a}+\mathbf{b}){\times}\mathbf{c} = \mathbf{a}{\times}\mathbf{c} + \mathbf{b}{\times}\mathbf{c}.$$

This proof, which is given purely for its physical interest, amounts to saying that the vector area of any closed surface is equal to zero. The relation holds, however nearly parallel to the plane of **a** and **b**, **c** may be. It may also be shown to hold when **c** lies in the plane of **a** and **b**. Conversely, assuming that the distributive law holds, the hydrostatic theorem employed in the above proof follows immediately.*

20. Cartesian Expansion for the Vector Product. It is often convenient to express a vector product in terms of the components of its vectors.

Let
$$\mathbf{a} = a_1\mathbf{i} + a_2\mathbf{j} + a_3\mathbf{k},$$
$$\mathbf{b} = b_1\mathbf{i} + b_2\mathbf{j} + b_3\mathbf{k};$$
then
$$\mathbf{a}{\times}\mathbf{b} = (a_1\mathbf{i} + a_2\mathbf{j} + a_3\mathbf{k}){\times}(b_1\mathbf{i} + b_2\mathbf{j} + b_3\mathbf{k})$$

which by the extension of (38) becomes

$$\mathbf{a}{\times}\mathbf{b} = (a_2b_3 - a_3b_2)\,\mathbf{i} + (a_3b_1 - a_1b_3)\,\mathbf{j} + (a_1b_2 - a_2b_1)\,\mathbf{k}. \qquad (39)$$

* Still another proof of the distributive law may be found in Föppl: Einführung in die Maxwell'sche Theorie der Elektricität, pp. 16 and 17.

This expression may be conveniently condensed into the determinant

$$\mathbf{a}\times\mathbf{b} = \begin{vmatrix} \mathbf{i} & \mathbf{j} & \mathbf{k} \\ a_1 & a_2 & a_3 \\ b_1 & b_2 & b_3 \end{vmatrix} \tag{40}$$

This is a useful mnemonic form for the vector product. As previously stated, if the vector product is zero the vectors, if finite, are parallel. This condition in terms of their projections on the axes is given by noticing that in (39) the three coefficients of **i**, **j**, and **k** must separately vanish, or again from the determinant form by noticing that two rows must be proportional, or that

$$\frac{a_1}{b_1} = \frac{a_2}{b_2} = \frac{a_3}{b_3},$$

a well known result.

If $\mathbf{a}_1$ and $\mathbf{b}_1$ are unit vectors $\mathbf{a}_1\times\mathbf{b}_1$ is the sine of their included angle θ; the quantities a_1, a_2, a_3, and b_1, b_2, b_3, being then their direction cosines respectively. Squaring formula (39) there results

$$\sin^2\theta = (a_2b_3 - a_3b_2)^2 + (a_3b_1 - a_1b_3)^2 + (a_1b_2 - a_2b_1)^2.$$

If we express the distributive law in the determinant form we obtain the following addition theorem in determinants of the third order.

$$\begin{vmatrix} \mathbf{i} & \mathbf{j} & \mathbf{k} \\ a_1 & a_2 & a_3 \\ (b_1 + c_1) & (b_2 + c_2) & (b_3 + c_3) \end{vmatrix} = \begin{vmatrix} \mathbf{i} & \mathbf{j} & \mathbf{k} \\ a_1 & a_2 & a_3 \\ b_1 & b_2 & b_3 \end{vmatrix} + \begin{vmatrix} \mathbf{i} & \mathbf{j} & \mathbf{k} \\ a_1 & a_2 & a_3 \\ c_1 & c_2 & c_3 \end{vmatrix} *$$

21. Application to Mechanics. Moment. The moment of a force **F** about a point O is defined as the product of the force into its perpendicular distance from the point O, or in symbols, by

$$F \times \overline{OA} = F \times r\sin\theta. \tag{41}$$

* Conversely, assuming the addition theorem for determinants, the distributive law of vector products follows immediately.

This moment, in the figure, is right-handed about a vector perpendicular to the paper and pointing directly up, so that a vector of magnitude $F \times \overline{OA}$ in this direction would represent the moment of **F** about O in a very convenient manner. According to this convention the vector

$$\mathbf{M} = \mathbf{r} \times \mathbf{F} \tag{42}$$

represents in *magnitude* and *direction* the moment of **F** about O, where **r** is the vector to the point of application of the force. If the force **F** is the resultant of a number of forces $\mathbf{F}_1$, $\mathbf{F}_2$ . . . acting at the same point of application, then by (38)

$$\mathbf{r} \times \mathbf{F} = \mathbf{r} \times (\mathbf{F}_1 + \mathbf{F}_2 + \ldots) = \mathbf{r} \times \mathbf{F}_1 + \mathbf{r} \times \mathbf{F}_2 + \ldots$$

or, the moment of the resultant of any number of forces about a point is equal to the sum of the separate moments. This theorem also shows that moments obey the parallelogram law. Conversely, assuming the truth of this theorem of moments, the distributive law for vector products is a necessary consequence.

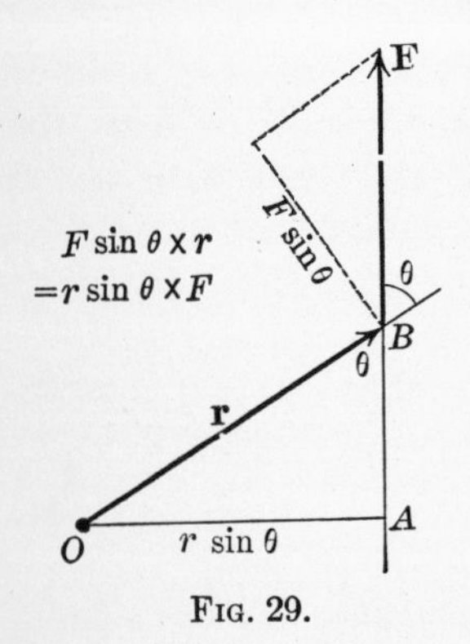

Fig. 29.

If **F** have components $X\ Y\ Z$, **r** components $x\ y\ z$, and **M** components $M_x\ M_y\ M_z$, the moment of **F** about the origin may be immediately written down by (40)

$$\mathbf{M} = \mathbf{r} \times \mathbf{F} = \begin{vmatrix} \mathbf{i} & \mathbf{j} & \mathbf{k} \\ x & y & z \\ X & Y & Z \end{vmatrix} = \mathbf{i}(yZ - zY) + \mathbf{j}(zX - xZ) + \mathbf{k}(xY - yX)$$

So that

$$\begin{aligned} M_x &= (yZ - zY), \\ M_y &= (zX - xZ), \\ M_z &= (xY - yX). \end{aligned} \tag{43}$$

22. Motion of a Rigid Body. Consider the motion of rotation of a rigid body about an axis, with a constant angular velocity $\boldsymbol{\omega}$. A velocity of rotation being of necessity about some axis, it is convenient to represent this kind of motion by a vector whose magnitude is proportional to the angular velocity and whose direction coincides with the axis of rotation. Its direction and that of the corresponding rotation may be simply represented by the symbol in Fig. 30.

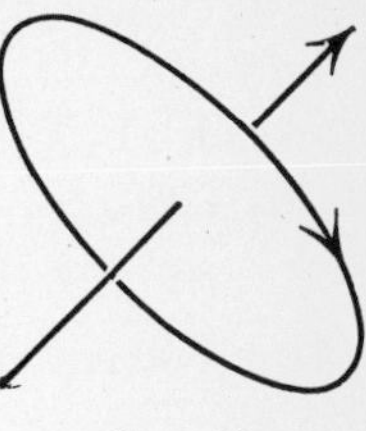

FIG. 30.

Notice that this is also the relation existing between direction of current and corresponding magnetic field.

Choose an origin on the axis of rotation Fig. 31 and consider a point P anywhere in the body, to find the velocity of the point P. Let P be determined by the radius vector $\mathbf{r}$ drawn to it from the origin. The velocity $\mathbf{q}$ of P is at right angles to $\boldsymbol{\omega}$ and to $\mathbf{r}$, its magnitude being given by the expression

$$q = \omega \times r \sin \theta,$$

as is easily seen in the figure. In other words $\mathbf{q}$ is represented not only in magnitude but *in direction* as well, by

$$\mathbf{q} = \boldsymbol{\omega}\times\mathbf{r}. \tag{44}$$

23. Composition of Angular Velocities. Since angular velocities may be represented by vectors let us see whether they compound according to the parallelogram law. To prove this definitely, let the body have several angular velocities $\boldsymbol{\omega}_1$, $\boldsymbol{\omega}_2$, $\boldsymbol{\omega}_3$ . . . about axes passing through the origin. Then the linear velocities of P separately due to these are

$$\mathbf{q}_1 = \boldsymbol{\omega}_1\times\mathbf{r},$$
$$\mathbf{q}_2 = \boldsymbol{\omega}_2\times\mathbf{r},$$
$$\mathbf{q}_3 = \boldsymbol{\omega}_3\times\mathbf{r},$$

and hence the velocity of P due to them all acting simultaneously is

$$\mathbf{q} = \mathbf{q}_1 + \mathbf{q}_2 + \mathbf{q}_3 + \cdots = \boldsymbol{\omega}_1 \times \mathbf{r} + \boldsymbol{\omega}_2 \times \mathbf{r} + \boldsymbol{\omega}_3 \times \mathbf{r} + \cdots$$
$$= (\boldsymbol{\omega}_1 + \boldsymbol{\omega}_2 + \boldsymbol{\omega}_3 + \cdots) \times \mathbf{r},$$

or the resultant velocity $\mathbf{q}$ of P is the same as if the body rotated with an angular velocity $\boldsymbol{\omega}$ about an axis through O given in magnitude and direction by

$$\boldsymbol{\omega} = \boldsymbol{\omega}_1 + \boldsymbol{\omega}_2 + \boldsymbol{\omega}_3 + \cdots = \Sigma \boldsymbol{\omega}.$$

This proves the above statement.

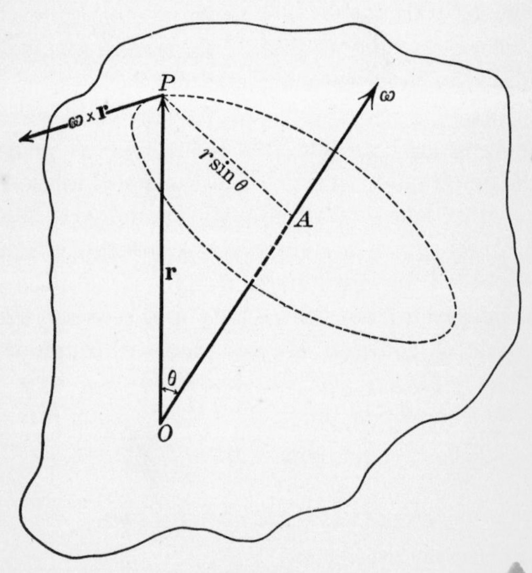

Fig. 31.

If the body have in addition to its angular velocity a velocity of translation $\mathbf{q}_t$ the resultant velocity $\mathbf{q}$ of the point P is simply

$$\mathbf{q} = \mathbf{q}_t + \boldsymbol{\omega} \times \mathbf{r}. \tag{45}$$

In the case that $\mathbf{q}_t$ is $\perp$ to $\boldsymbol{\omega}$, there must be a line of points which are instantaneously at rest. This line is determined by the condition $\mathbf{q} = 0$ or

$$\mathbf{r} \times \boldsymbol{\omega} = \mathbf{q}_t, \tag{46}$$

which is a straight line parallel to $\boldsymbol{\omega}$. Change the origin to a point O' on this line, the expression for the velocity reduces to the form

$$\mathbf{q} = \boldsymbol{\omega} \times \mathbf{r}',$$

where $\mathbf{r}'$ is the vector from O' to any point in the body. If $\mathbf{q}_t$ is not $\perp$ to $\boldsymbol{\omega}$, decompose it into two components $\mathbf{q}_t'$ and $\mathbf{q}_t''$ such that $\mathbf{q}_t = \mathbf{q}_t' + \mathbf{q}_t''$.

Let $\mathbf{q}_t'$ be $\parallel$ to $\boldsymbol{\omega}$, and $\mathbf{q}_t''$ $\perp$ to $\boldsymbol{\omega}$, we may then proceed as before with $\mathbf{q}_t''$. It is thus seen that the most general motion possible of a rigid body is that of a rotation about a certain axis and a velocity of translation along it; in other words, a screw motion.

If $\boldsymbol{\omega}$ and $\mathbf{q}_t$ are variable this holds true at any instant, although the direction and pitch of the screw motion may be rapidly and of course continuously changing. The axis of rotation about which a rigid body is rotating at any given instant is called the **instantaneous axis** of rotation. If the body has one point fixed the velocity $\mathbf{q}_t$ is zero and the instantaneous axis of rotation always passes through this fixed point. The equation of the instantaneous axis is then given by the condition that

$$\mathbf{r} \times \boldsymbol{\omega} = \mathbf{q}_t''.$$

EXERCISES AND PROBLEMS.

1. Show that the two vectors

$$\mathbf{a} = 9\,\mathbf{i} + \mathbf{j} - 6\,\mathbf{k}$$

and

$$\mathbf{b} = 4\,\mathbf{i} - 6\,\mathbf{j} + 5\,\mathbf{k}$$

are at right angles to each other.

2. The coördinates of two points are (3, 1, 2) and (2, – 2, 4); find the cosine of the angle between the vectors joining these points to the origin.

3. Write out in the form

$$\mathbf{a} = a_1\mathbf{i} + a_2\mathbf{j} + a_3\mathbf{k}$$

several pairs of mutually perpendicular vectors.

4. Write out in the form

$$\mathbf{a} = a_1\mathbf{i} + a_2\mathbf{j} + a_3\mathbf{k}$$

the expressions for several unit vectors.

5. Find a vector in the **ij**-plane which has the same length as the vector

$$\mathbf{a} = 4\,\mathbf{i} - 2\,\mathbf{j} + 3\,\mathbf{k}.$$

Find a vector in the **jk**-plane having the same length, and the same **j**-projection as **a**.

6. Let **a** and **b** be two unit vectors lying in the **ij**-plane. Let α be the angle that **a** makes with **i**, and β the angle **b** makes with **i**; then

$$\mathbf{a} = \mathbf{i}\cos\alpha + \mathbf{j}\sin\alpha,$$

$$\mathbf{b} = \mathbf{i}\cos\beta + \mathbf{j}\sin\beta.$$

Form the dot and cross products and show that the addition theorems for the cosine and sine follow from their interpretation.

7. Let

$$\mathbf{a} = a_1\mathbf{i} + a_2\mathbf{j} + a_3\mathbf{k}$$

and

$$\mathbf{b} = b_1\mathbf{i} + b_2\mathbf{j} + b_3\mathbf{k}$$

be the unit vectors to the points A and B. Find the distance between A and B and its direction cosines in terms of a_1, a_2, a_3 and b_1, b_2, b_3.

8. Three vectors of lengths a, $2\,a$, $3\,a$ meet in a point and are directed along the diagonals of the three faces of a cube, meeting at the point. Determine the magnitude of their resultant. Find the resultant in the form

$$\mathbf{r} = x\,\mathbf{i} + y\,\mathbf{j} + z\,\mathbf{k}$$

and from this calculate its magnitude.

9. The sum of the squares of the diagonals of a parallelogram is equal to the sum of the squares of the sides.

10. Parallelograms upon the same base and between same parallels are equal in area.

11. The squares of the sides of *any* quadrilateral exceed the squares of the diagonals by four times the square of the line which joins the middle points of the diagonals.

12. Under what conditions will the resultant of a system of vectors of magnitudes 7, 24, and 25 be equal to zero?

13. Three vectors of lengths a, a, and $a\sqrt{2}$ meet in a point and are mutually at right angles. Determine the magnitude of the resultant and the angles between its direction and that of each component.

14. ABC is a triangle, and P any point in BC. If PQ represent the resultant of the forces represented by AP, PB, BC, show that the locus of Q is a straight line parallel to BC.

15. The angle in a semicircle is a right angle.

Take equation of circle

$$\mathbf{r}^2 - 2\,\mathbf{a}\cdot\mathbf{r} = 0,$$

factor with **r** and interpret.

16. If two circles intersect, the line joining the centers is perpendicular to the line joining the points of intersection.

17. O is a fixed point, AB a given straight line. A point Q is taken in the line OP drawn to a point P in AB, such that

$$OP \cdot OQ = k^2 \text{ (a const.)}.$$

To find the locus of Q.

Application to problems in Inversion.

18. If any line pass through the centroid of a number of points, the sum of the perpendiculars on this line from the different points, measured in the same direction, is zero.

Application to method of Least Squares.

19. Write out the vector product of the two vectors

$$\mathbf{a} = 6\,\mathbf{i} + 0.3\,\mathbf{j} - 5\,\mathbf{k}$$

and

$$\mathbf{b} = 0.1\,\mathbf{i} - 4.2\,\mathbf{j} + 2.5\,\mathbf{k}$$

and show by calculation that the resulting vector is perpendicular to each of the constituent vectors of the product.

20. Find the area of the triangle determined by the two vectors

$$\mathbf{a} = 3\,\mathbf{i} + 4\,\mathbf{j}$$

and

$$\mathbf{b} = -5\,\mathbf{i} + 7\,\mathbf{j}.$$

21. Find the area of the parallelogram determined by the vectors

$$\mathbf{a} = \mathbf{i} + 2\,\mathbf{j} + 3\,\mathbf{k}$$

and

$$\mathbf{b} = -3\,\mathbf{i} - 2\,\mathbf{j} + \mathbf{k}.$$

22. Express the relations between the sides and opposite angles of a triangle.

In any triangle of vector sides **a**, **b**, **c**,

$$\mathbf{a} = \mathbf{b} - \mathbf{c},$$

take the vector product of **a** with this and interpret.

23. By means of the equation of § 20 find the sine of the angle between the two vectors

$$\mathbf{a} = 3\,\mathbf{i} + \mathbf{j} + 2\,\mathbf{k}$$

and

$$\mathbf{b} = 2\,\mathbf{i} - 2\,\mathbf{j} + 4\,\mathbf{k}.$$

24. Show that the equation of a line perpendicular to the two vectors **b** and **c** is

$$\mathbf{r} = \mathbf{a} + x\,\mathbf{b}\times\mathbf{c}.$$

25. Find the perpendicular from the origin on the line

$$\mathbf{a}\times(\mathbf{r} - \mathbf{b}) = 0.$$

26. Derive an expression for the area of a square of which

$$\mathbf{r} = a_1\mathbf{i} + a_2\mathbf{j}$$

is the semi-diagonal.

27. If the middle point of one of the non-parallel sides of a trapezoid be joined to the extremities of the opposite side, a triangle is obtained whose area is one-half of that of the trapezoid.

28. Find the relations between two right-handed systems of three mutually perpendicular unit vectors. See Gibbs-Wilson, p. 104.

29. Given $\mathbf{c} = \mathbf{a} + \mathbf{b}.$

Expand the right-hand side of each of the equations

$$\mathbf{c}\cdot\mathbf{c} = (\mathbf{a} + \mathbf{b})\cdot\mathbf{c},$$
$$\mathbf{c}\cdot\mathbf{c} = (\mathbf{a} + \mathbf{b})\cdot(\mathbf{a} + \mathbf{b}),$$

and give the geometric interpretation of the result.

30. Given $$\mathbf{r} = x\,\mathbf{a} + y\,\mathbf{b} + z\,\mathbf{c}$$
where $\mathbf{a}\ \mathbf{b}\ \mathbf{c}$ are three non-coplanar vectors. Expand the right-hand side of the equation
$$\mathbf{r}\cdot\mathbf{r} = (x\,\mathbf{a} + y\,\mathbf{b} + z\,\mathbf{c})\cdot(x\,\mathbf{a} + y\,\mathbf{b} + z\,\mathbf{c})$$
and give the geometric interpretation of the result.

31. Show that the work done by a force during a displacement is equal to the sum of the quantities of work performed by its components during the displacement.

32. A fluid is flowing across a plane surface with a uniform velocity which is represented in magnitude and direction by the vector $\mathbf{q}$. If $\mathbf{n}$ is the unit normal to the plane, show that the volume of the fluid that passes through the unit area of the plane in unit time is $\mathbf{q}\cdot\mathbf{n}$.

33. Show that a system of forces represented in magnitude, direction, and position by the successive sides of a plane polygon is equivalent to a couple whose moment is equal to twice the area of the polygon.

34. If O be any point whatever, either in the plane of the triangle ABC or out of that plane, the squares of the sides of the triangle fall short of three times the squares of the distances of the angular points from O, by the square of three times the distance of the mean point from O.

35. The sum of the squares of the distances of any point O from the angular points of the triangle exceeds the sum of the squares of its distances from the middle points of the sides by the sum of the squares of half the sides.

36. Show that
$$(\mathbf{a} - \mathbf{b})\times(\mathbf{a} + \mathbf{b}) = 2\,\mathbf{a}\times\mathbf{b}.$$
and give its geometric interpretation,

37. Show that
$$(\mathbf{a} - \mathbf{b})\cdot(\mathbf{a} + \mathbf{b}) = \mathbf{a}^2 - \mathbf{b}^2$$
and interpret.

CHAPTER III.

VECTOR AND SCALAR PRODUCTS INVOLVING THREE VECTORS.

24. From the three vectors a, b, and c the following combinations may be derived:

1. $\mathbf{a}\,(\mathbf{b}\cdot\mathbf{c})$ (a vector)	4. $\mathbf{a}\,(\mathbf{b}\times\mathbf{c})$ (not defined)	
2. $\mathbf{a}\cdot(\mathbf{b}\times\mathbf{c})$ (a scalar)	5. $\mathbf{a}\cdot(\mathbf{b}\cdot\mathbf{c})$ (absurd)	(47)
3. $\mathbf{a}\times(\mathbf{b}\times\mathbf{c})$ (a vector)	6. $\mathbf{a}\times(\mathbf{b}\cdot\mathbf{c})$ (absurd).	

Of these six expressions, 5 and 6 are meaningless and absurd, because they are the scalar product and vector product, respectively, of a *vector* ($\mathbf{a}$) and a *scalar* ($\mathbf{b}\cdot\mathbf{c}$), and such products require a vector on *each* side of the dot or cross. As to 4, since no definition of the product of two vectors without a dot or a cross has been made, it is as yet meaningless. In this book we shall not consider such products. We shall consider in detail the three remaining triple products. The first one of these, $\mathbf{a}\,(\mathbf{b}\cdot\mathbf{c})$, is simply the vector $\mathbf{a}$ multiplied by the scalar quantity ($\mathbf{b}\cdot\mathbf{c}$) and is a vector in the same direction as $\mathbf{a}$, but $bc \cos (\mathbf{bc})$ times longer. This triple product, then, offers no new difficulties, and means

$$\mathbf{a}\,(\mathbf{b}\cdot\mathbf{c}) = \mathbf{a} \times bc \cos (\mathbf{bc}).$$

25. The Triple Product $V = \mathbf{a}\cdot(\mathbf{b}\times\mathbf{c})$ is a scalar and represents the volume of a parallelopiped of which the three conterminous edges are $\mathbf{a}$, $\mathbf{b}$, and $\mathbf{c}$. This is easily seen to be the case, as $\mathbf{b}\times\mathbf{c}$ is the area of the base represented by a vector $\overline{OS} \perp$ to this base; the scalar product of $\mathbf{a}$ and the vector $\overline{OS}$ will be this area multiplied by the projection of the slant height $\mathbf{a}$ along it, or, in other words, the volume. As evidently this volume, V, may be obtained by forming the vector products of any two of the three vectors **a**, **b**, and **c** (thus giving the area of one of the faces) and forming the scalar product of this vector-area with the remaining third vector, it follows that

$$V = \mathbf{b}\cdot(\mathbf{c}\times\mathbf{a}) = \mathbf{c}\cdot(\mathbf{a}\times\mathbf{b}) = \mathbf{a}\cdot(\mathbf{b}\times\mathbf{c}).$$

If the vectors $(\mathbf{c}\times\mathbf{a})$, $(\mathbf{a}\times\mathbf{b})$, and $(\mathbf{b}\times\mathbf{c})$ are taken so that they form an acute angle with **b**, **c**, and **a**, respectively, then the volume is to be considered positive, the cosine term in the

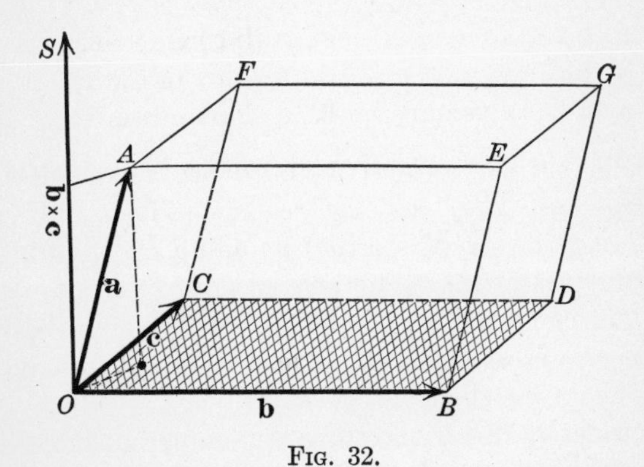

FIG. 32.

scalar product being positive. Otherwise the volume is to be considered negative. Of course the inversion of the factors in the vector products should change the sign, by (33), so that we have

$$\begin{aligned} V \equiv \mathbf{b}\cdot(\mathbf{c}\times\mathbf{a}) &= (\mathbf{c}\times\mathbf{a})\cdot\mathbf{b} = -\,\mathbf{b}\cdot(\mathbf{a}\times\mathbf{c}) = -\,(\mathbf{a}\times\mathbf{c})\cdot\mathbf{b} \\ = \mathbf{c}\cdot(\mathbf{a}\times\mathbf{b}) &= (\mathbf{a}\times\mathbf{b})\cdot\mathbf{c} = -\,\mathbf{c}\cdot(\mathbf{b}\times\mathbf{a}) = -\,(\mathbf{b}\times\mathbf{a})\cdot\mathbf{c} \qquad (48) \\ = \mathbf{a}\cdot(\mathbf{b}\times\mathbf{c}) &= (\mathbf{b}\times\mathbf{c})\cdot\mathbf{a} = -\,\mathbf{a}\cdot(\mathbf{c}\times\mathbf{b}) = -\,(\mathbf{c}\times\mathbf{b})\cdot\mathbf{a}. \end{aligned}$$

By a consideration of these equalities the following laws may be seen to hold:

1. The sign of the scalar triple product is unchanged as long as the cyclical order of the factors is unchanged.

2. For every change of cyclical order a minus sign is introduced.

3. The dot and the cross may be interchanged *ad libitum.*

The equalities (48) are called by **Heaviside** the **Parallelopiped Law.**

The product $\mathbf{a}\cdot(\mathbf{b}\times\mathbf{c})$ may be written in terms of the components of its vectors along any three rectangular Cartesian axes as

$$\begin{aligned}\mathbf{a}\cdot(\mathbf{b}\times\mathbf{c}) &= a_1(b_2c_3 - b_3c_2) + a_2(b_3c_1 - b_1c_3) + a_3(b_1c_2 - b_2c_1)\\ &= a_1\begin{vmatrix} b_2 & b_3 \\ c_2 & c_3\end{vmatrix} + a_2\begin{vmatrix} b_3 & b_1 \\ c_3 & c_1\end{vmatrix} + a_3\begin{vmatrix} b_1 & b_2 \\ c_1 & c_2\end{vmatrix}\\ &= \begin{vmatrix} a_1 & a_2 & a_3 \\ b_1 & b_2 & b_3 \\ c_1 & c_2 & c_3\end{vmatrix}\end{aligned} \tag{49}$$

This is the familiar determinant expression for the volume of a parallelopiped with one corner at the origin.

The parallelopiped principle, then, expresses the fact that as long as the cyclical order of the rows is unchanged the determinant is also unchanged, but that every interchange of cyclical order introduces a minus sign as a factor. To the student familiar with determinants this is a well known property. Conversely, assuming this property of a determinant as proven, the equations (48) immediately follow.

The twelve expressions (48) are often written in one, as $[\mathbf{abc}]$, a special symbol of abbreviation taken from Grassmann.

26. Condition that Three Vectors Lie in One Plane. Should the three finite non-parallel vectors $\mathbf{a}$, $\mathbf{b}$, and $\mathbf{c}$ lie in one plane the volume of the parallelopiped they determine is zero. Hence the condition that the three non-parallel vectors should lie in a plane is that

$$[\mathbf{abc}] = 0. \tag{50}$$

In the expression $[\mathbf{abc}]$, if any two of the vectors are parallel the volume of the parallelopiped is again evidently zero. Hence, in general, $$[\mathbf{aab}] = 0. \tag{51}$$

To look at it in another way, we may put, by (48)

$$\mathbf{a}\cdot(\mathbf{a}\times\mathbf{b}) = (\mathbf{a}\times\mathbf{a})\cdot\mathbf{b},$$

and as $\mathbf{a}\times\mathbf{a} = 0$, then $\mathbf{a}\cdot(\mathbf{a}\times\mathbf{b}) = 0$, so that in a triple scalar product if *any* two of the vectors are parallel their triple scalar product is zero.

In the determinant (49) above, this corresponds to having any two rows proportional to each other, the result being, as is well known, identically zero.

The parenthesis in an expression such as $\mathbf{a}\cdot(\mathbf{b}\times\mathbf{c})$ is in reality unnecessary, as its only other interpretation $(\mathbf{a}\cdot\mathbf{b})\times\mathbf{c}$ is without meaning, being the vector product of a scalar $(\mathbf{a}\cdot\mathbf{b})$ and a vector $\mathbf{c}$. The parentheses are introduced, however, when by so doing the interpretation is made easier.

Scalar magnitudes of the vectors, it is important to remember, which occur in any kind of scalar or vector products may be placed in *any part* of the expression as factors.

For example,

$$\begin{aligned} \mathbf{a}\cdot(\mathbf{b}\times\mathbf{c}) &= a\mathbf{a}_1\cdot(\mathbf{b}\times\mathbf{c}) \\ &= abc\,\mathbf{a}_1\cdot(\mathbf{b}_1\times\mathbf{c}_1) \qquad (52) \\ &= b\mathbf{a}_1\cdot(c\mathbf{b}_1\times a\mathbf{c}_1), \text{ etc., etc.} \end{aligned}$$

27. The Triple Product $\mathbf{q} = \mathbf{a}\times(\mathbf{b}\times\mathbf{c})$ is a vector. In this expression the parenthesis, or some separating symbol, is necessary, as $\mathbf{a}\times(\mathbf{b}\times\mathbf{c}) \neq (\mathbf{a}\times\mathbf{b})\times\mathbf{c}$. The sign of this product changes every time the order of the factors $\mathbf{a}$ and $(\mathbf{b}\times\mathbf{c})$ is changed in $\mathbf{a}\times(\mathbf{b}\times\mathbf{c})$, or whenever the order of the factors $\mathbf{b}$ and $\mathbf{c}$ is changed in $(\mathbf{b}\times\mathbf{c})$. The vector product being always perpendicular to both of its components, $\mathbf{q}$ is perpendicular to $\mathbf{a}$ as well as to $\mathbf{b}\times\mathbf{c}$, hence

$$\mathbf{q}\cdot\mathbf{a} = 0 \text{ and } \mathbf{q}\cdot(\mathbf{b}\times\mathbf{c}) = 0. \tag{53}$$

Equation (53) shows that $\mathbf{q}$ lies in the same plane as $\mathbf{b}$ and $\mathbf{c}$, either by (50) or by seeing that it is perpendicular to a line which is itself perpendicular to $\mathbf{b}$ and $\mathbf{c}$. It is important that this result be clearly visualized. The habit of visualization should be cultivated, as it is of great importance to the student whatever kind of analysis he be using, but particularly so in this. To a *purely* analytical mind vector analysis offers but few advantages.

As $\mathbf{q}$ lies in the plane of $\mathbf{b}$ and $\mathbf{c}$ it is possible to express $\mathbf{q}$ in the form

$$\mathbf{q} = x\,\mathbf{b} - y\,\mathbf{c},$$

where x and y are scalar multipliers. Let us try to determine the quantities x and y. Since $\mathbf{q}$ is perpendicular to $\mathbf{a}$,

$$\mathbf{a}\cdot\mathbf{q} = x\,\mathbf{a}\cdot\mathbf{b} - y\,\mathbf{a}\cdot\mathbf{c} = 0$$

and, therefore,

$$x : y = \mathbf{a}\cdot\mathbf{c} : \mathbf{a}\cdot\mathbf{b} \quad \text{or } x = n\,\mathbf{a}\cdot\mathbf{c},$$
$$y = n\,\mathbf{a}\cdot\mathbf{b},$$

where n is a scalar factor of proportionality. So that

$$\mathbf{q} = \mathbf{a}\times(\mathbf{b}\times\mathbf{c}) = n\,[\mathbf{b}(\mathbf{a}\cdot\mathbf{c}) - \mathbf{c}(\mathbf{a}\cdot\mathbf{b})]. \tag{54}$$

We shall now prove that n is independent of the magnitudes and inclinations of the vectors $\mathbf{a}$, $\mathbf{b}$, and $\mathbf{c}$. It is independent of their magnitudes because they may be taken out by (52) as scalar coefficients and eliminated from the equation. Since we are dealing with the mutual relations between any three vectors, we may choose one of them arbitrarily. Let that one be $\mathbf{a}$. Let us now replace one of the remaining vectors, $\mathbf{c}$, for instance, by the sum of two other vectors $\mathbf{d}$ and $\mathbf{e}$. Then

$$\mathbf{a}\times(\mathbf{b}\times(\mathbf{d} + \mathbf{e})) = n\,[\mathbf{b}\;\mathbf{a}\cdot(\mathbf{d} + \mathbf{e}) - (\mathbf{d} + \mathbf{e})\;\mathbf{a}\cdot\mathbf{b}],$$

or

$$\mathbf{a}\times(\mathbf{b}\times\mathbf{d}) + \mathbf{a}\times(\mathbf{b}\times\mathbf{e}) = n\,[\mathbf{b}\;\mathbf{a}\cdot(\mathbf{d} + \mathbf{e}) - (\mathbf{d}+\mathbf{e})\;\mathbf{a}\cdot\mathbf{b}],$$

or finally,

$$n'[\mathbf{b}\;\mathbf{a}\cdot\mathbf{d} - \mathbf{d}\;\mathbf{a}\cdot\mathbf{b}] + n''[\mathbf{b}\;\mathbf{a}\cdot\mathbf{e} - \mathbf{e}\;\mathbf{a}\cdot\mathbf{b}] = n[\mathbf{b}\;\mathbf{a}\cdot(\mathbf{d}+\mathbf{e}) - (\mathbf{d}+\mathbf{e})\,\mathbf{a}\cdot\mathbf{b}].$$

If **d** and **e** have been chosen so that **b**, **d**, and **e** are not coplanar, then we may equate coefficients of the vectors **b**, **d**, and **e** on both sides. Thus

$$n'\mathbf{a}\cdot\mathbf{d} + n''\,\mathbf{a}\cdot\mathbf{e} = n\,\mathbf{a}\cdot(\mathbf{d} + \mathbf{e}).$$

$$n'\,\mathbf{a}\cdot\mathbf{b} = n\,\mathbf{a}\cdot\mathbf{b},$$

$$n''\,\mathbf{a}\cdot\mathbf{b} = n\,\mathbf{a}\cdot\mathbf{b},$$

which necessitates that

$$n' = n'' = n.$$

The coefficient n in equation (54) is thus independent of **a**, **b**, and **c**, and is, therefore, a numerical constant. To find its value we are now at liberty to consider a special case. Let **a**, **b**, **c** be unit vectors. Let $\mathbf{a} = \mathbf{c}$ and let **b** be perpendicular to **c**. This is the equivalent to writing $\mathbf{a} = \mathbf{k}$, $\mathbf{b} = \mathbf{j}$, $\mathbf{c} = \mathbf{k}$. We then have for equation (54),

$$\mathbf{k}\times(\mathbf{j}\times\mathbf{k}) = n\,[\mathbf{j}(\mathbf{k}\cdot\mathbf{k}) - \mathbf{k}(\mathbf{k}\cdot\mathbf{j})].$$

but

$$\mathbf{j}\times\mathbf{k} = \mathbf{i} \text{ and } \mathbf{k}\times\mathbf{i} = \mathbf{j},$$

$$\mathbf{k}\cdot\mathbf{k} = 1 \text{ and } \mathbf{k}\cdot\mathbf{j} = 0:$$

therefore the equation reduces to

$$\mathbf{j} = n(\mathbf{j}); \qquad \text{hence } n = 1.$$

We have thus proved the very important relation

$$\mathbf{a}\times(\mathbf{b}\times\mathbf{c}) = \mathbf{b}(\mathbf{a}\cdot\mathbf{c}) - \mathbf{c}(\mathbf{a}\cdot\mathbf{b}), \tag{55}$$

which should be memorized.

28. Demonstration by Cartesian Expansion. A demonstration of this equation may also be obtained by expanding in terms of the Cartesian components of the vectors. This method is a very useful one when no other demonstration readily offers itself, but generally (not in this case) has the disadvantage of being long and cumbersome. No better examples of the concentration of the vector notation may be found than by carrying through a number of such transformations. On account of the importance of the equation (55), and also to give an example of the expansion method in general, its demonstration by this method will be carried out.

As the components of $(\mathbf{b}\times\mathbf{c})$ are $\begin{vmatrix} b_2 b_3 \\ c_2 c_3 \end{vmatrix}$, $\begin{vmatrix} b_3 b_1 \\ c_3 c_1 \end{vmatrix}$ and $\begin{vmatrix} b_1 b_2 \\ c_1 c_2 \end{vmatrix}$, see (49) we may write, by (40),

$$\mathbf{a}\times(\mathbf{b}\times\mathbf{c}) = \begin{vmatrix} \mathbf{i} & \mathbf{j} & \mathbf{k} \\ a_1 & a_2 & a_3 \\ \begin{vmatrix} b_2 b_3 \\ c_2 c_3 \end{vmatrix} & \begin{vmatrix} b_3 b_1 \\ c_3 c_1 \end{vmatrix} & \begin{vmatrix} b_1 b_2 \\ c_1 c_2 \end{vmatrix} \end{vmatrix} \begin{array}{l} = \mathbf{i}\,[a_2(b_1c_2 - b_2c_1) - a_3(b_3c_1 - b_1c_3)] \\ + \mathbf{j}\,[a_3(b_2c_3 - b_3c_2) - a_1(b_1c_2 - b_2c_1)] \\ + \mathbf{k}\,[a_1(b_3c_1 - b_1c_3) - a_2(b_2c_3 - b_3c_2)]. \end{array}$$

The terms may now be rearranged into

$$\begin{aligned} \mathbf{a}\times(\mathbf{b}\times\mathbf{c}) = {} & \mathbf{i}\,b_1\,(\underline{a_1c_1} + a_2c_2 + a_3c_3) \\ & + \mathbf{j}\,b_2\,(a_1c_1 + \underline{a_2c_2} + a_3c_3) \\ & + \mathbf{k}\,b_3\,(a_1c_1 + a_2c_2 + \underline{a_3c_3}) \\ & - \mathbf{i}\,c_1\,(\underline{a_1b_1} + a_2b_2 + a_3b_3) \\ & - \mathbf{j}\,c_2\,(a_1b_1 + \underline{a_2b_2} + a_3b_3) \\ & - \mathbf{k}\,c_3\,(a_1b_1 + a_2b_2 + \underline{a_3b_3}). \end{aligned}$$

The new underlined terms have been added and subtracted. The first three lines are

$$(\mathbf{i}\,b_1 + \mathbf{j}\,b_2 + \mathbf{k}\,b_3)\,(\mathbf{a}\cdot\mathbf{c}) = \mathbf{b}(\mathbf{a}\cdot\mathbf{c}),$$

the last three are

$$-(\mathbf{i}\,c_1 + \mathbf{j}\,c_2 + \mathbf{k}\,c_3)\,(\mathbf{a}\cdot\mathbf{b}) = -\,\mathbf{c}(\mathbf{a}\cdot\mathbf{b}),$$

hence

$$\mathbf{a}\times(\mathbf{b}\times\mathbf{c}) = \mathbf{b}(\mathbf{a}\cdot\mathbf{c}) - \mathbf{c}(\mathbf{a}\cdot\mathbf{b}).$$

29. Third Proof. That $n = 1$ in (54) may also be proved as follows: Consider first the triple vector product in which two of the vectors are the same,

$$\mathbf{b}\times(\mathbf{b}\times\mathbf{c}) = n\,(\mathbf{b}\;\mathbf{b}\cdot\mathbf{c} - \mathbf{c}\;\mathbf{b}\cdot\mathbf{b}).$$

Taking the scalar product of this and **c**, or, in shorter language, applying **c** dot (**c**·) to it, we obtain

$$\mathbf{c}\cdot\mathbf{b}\times(\mathbf{b}\times\mathbf{c}) = n[(\mathbf{b}\cdot\mathbf{c})^2 - \mathbf{b}^2\mathbf{c}^2].$$

But by an interchange of the dot and the cross and one change of cyclical order the left-hand side becomes

$$(\mathbf{c}\times\mathbf{b})\cdot(\mathbf{b}\times\mathbf{c}) = -(\mathbf{b}\times\mathbf{c})\cdot(\mathbf{b}\times\mathbf{c}) = -(\mathbf{b}\times\mathbf{c})^2. \qquad (56)$$

We know, however, that

$$(\mathbf{b}\cdot\mathbf{c})^2 + (\mathbf{b}\times\mathbf{c})^2 = \mathbf{b}^2\mathbf{c}^2, \qquad (56a)$$

as by definition

$$b^2c^2 \cos^2 (\mathbf{bc}) + b^2c^2 \sin^2 (\mathbf{bc}) = b^2c^2$$

is equivalent to it; hence the right hand of the first equation is nothing more than $-n\,(\mathbf{b}\times\mathbf{c})^2$, and comparing with (56) we see that n must be unity. The theorem is thus true, when two of the vectors are the same. Consider now the general case,

$$\mathbf{a}\times(\mathbf{b}\times\mathbf{c}) = n\,(\mathbf{b}\;\mathbf{a}\cdot\mathbf{c} - \mathbf{c}\;\mathbf{a}\cdot\mathbf{b}). \qquad (57)$$

Apply **b** dot to it, obtaining

$$\mathbf{b}\cdot\mathbf{a}\times(\mathbf{b}\times\mathbf{c}) = n\,(\mathbf{b}\cdot\mathbf{b}\;\mathbf{a}\cdot\mathbf{c} - \mathbf{b}\cdot\mathbf{c}\;\mathbf{a}\cdot\mathbf{b})$$
$$= n\,\mathbf{a}\cdot(\mathbf{c}\;\mathbf{b}\cdot\mathbf{b} - \mathbf{b}\;\mathbf{b}\cdot\mathbf{c}),$$

which as we have just proved may be written

$$= n\,\mathbf{a}\cdot[-\,\mathbf{b}\times(\mathbf{b}\times\mathbf{c})].$$

But on the left-hand side we have

$$\mathbf{b}\cdot\mathbf{a}\times(\mathbf{b}\times\mathbf{c}) = -\,\mathbf{a}\cdot[\mathbf{b}\times(\mathbf{b}\times\mathbf{c})],$$

by an interchange of dot and cross and one of cyclical order. Comparing the last two equations we see that $n = 1$ in general.

The parenthesis in $\mathbf{a}\times(\mathbf{b}\times\mathbf{c})$ is necessary, for $(\mathbf{a}\times\mathbf{b})\times\mathbf{c}$ is quite different from the first expression, as one may readily see by expanding the two, or by the reasoning of § 27.

30. Products of More than Three Vectors. In practical applications to physics more complicated products than those of three vectors seldom arise. Whenever they do, they may be reduced by successive applications of the preceding principles. In any case they represent *extremely* complicated Cartesian expansions. As an example of such a reduction, consider the scalar expression containing four vectors, $(\mathbf{a}\times\mathbf{b})\cdot(\mathbf{c}\times\mathbf{d})$.

Interchange the first cross and dot and expand the vector triple product, which will give,

$$\begin{aligned}(\mathbf{a}\times\mathbf{b})\cdot(\mathbf{c}\times\mathbf{d}) &= \mathbf{a}\cdot\mathbf{b}\times(\mathbf{c}\times\mathbf{d}) \\ &= \mathbf{a}\cdot(\mathbf{c}\ \mathbf{b}\cdot\mathbf{d} - \mathbf{d}\ \mathbf{b}\cdot\mathbf{c}) \\ &= \mathbf{a}\cdot\mathbf{c}\ \mathbf{b}\cdot\mathbf{d} - \mathbf{a}\cdot\mathbf{d}\ \mathbf{b}\cdot\mathbf{c} \\ &= \begin{vmatrix} \mathbf{a}\cdot\mathbf{c} & \mathbf{b}\cdot\mathbf{c} \\ \mathbf{a}\cdot\mathbf{d} & \mathbf{b}\cdot\mathbf{d} \end{vmatrix}. \end{aligned} \tag{58}$$

This formula will be used in the deduction of Stokes' Theorem in § (58).

Again consider the quadruple vector product $(\mathbf{a}\times\mathbf{b})\times(\mathbf{c}\times\mathbf{d})$, which may be expanded by (55),

$$(\mathbf{a}\times\mathbf{b})\times(\mathbf{c}\times\mathbf{d}) = \mathbf{c}\ (\mathbf{a}\times\mathbf{b})\cdot\mathbf{d} - \mathbf{d}\ (\mathbf{a}\times\mathbf{b})\cdot\mathbf{c}$$

or into

$$= \mathbf{b}\ (\mathbf{c}\times\mathbf{d})\cdot\mathbf{a} - \mathbf{a}\ (\mathbf{c}\times\mathbf{d})\cdot\mathbf{b}, \tag{59}$$

taking in the first case $(\mathbf{a}\times\mathbf{b})$, $\mathbf{c}$, and $\mathbf{d}$ as the three vectors of a triple product, and in the second case $\mathbf{a}$, $\mathbf{b}$, and $(\mathbf{c}\times\mathbf{d})$. By subtracting these two equal expressions from each other we have

$$\mathbf{a}\ \mathbf{b}\cdot(\mathbf{c}\times\mathbf{d}) - \mathbf{b}\ \mathbf{a}\cdot(\mathbf{c}\times\mathbf{d}) + \mathbf{c}\ \mathbf{d}\cdot(\mathbf{a}\times\mathbf{b}) - \mathbf{d}\ \mathbf{c}\cdot(\mathbf{a}\times\mathbf{b}) \equiv 0, \tag{60}$$

an important relation holding between *any* four vectors. Putting $\mathbf{d} = \mathbf{r}$, this equation may be written

$$\mathbf{r}\ [\mathbf{abc}] = [\mathbf{rbc}]\mathbf{a} + [\mathbf{rca}]\mathbf{b} + [\mathbf{rab}]\mathbf{c},$$

so that

$$\mathbf{r} = \frac{[\mathbf{rbc}]}{[\mathbf{abc}]}\,\mathbf{a} + \frac{[\mathbf{rca}]}{[\mathbf{abc}]}\,\mathbf{b} + \frac{[\mathbf{rab}]}{[\mathbf{abc}]}\,\mathbf{c}, \tag{61}$$

or

$$\mathbf{r} = \mathbf{r}\cdot\frac{\mathbf{b}\times\mathbf{c}}{[\mathbf{abc}]}\,\mathbf{a} + \mathbf{r}\cdot\frac{\mathbf{c}\times\mathbf{a}}{[\mathbf{abc}]}\,\mathbf{b} + \mathbf{r}\cdot\frac{\mathbf{a}\times\mathbf{b}}{[\mathbf{abc}]}\,\mathbf{c}, \tag{62}$$

an important and useful formula which gives the coefficients necessary to express $\mathbf{r}$ in terms of any three arbitrary vectors not lying in the same plane. This expansion is under these conditions always possible as explained in § (7).

31. Reciprocal System of Vectors. The three vectors

$$\frac{\mathbf{b}\times\mathbf{c}}{[\mathbf{abc}]}, \quad \frac{\mathbf{c}\times\mathbf{a}}{[\mathbf{abc}]}, \quad \text{and} \quad \frac{\mathbf{a}\times\mathbf{b}}{[\mathbf{abc}]}, \tag{63}$$

perpendicular respectively to the planes of $\mathbf{b}$ and $\mathbf{c}$, $\mathbf{c}$ and $\mathbf{a}$, and $\mathbf{a}$ and $\mathbf{b}$, occur frequently in important relations and are said to be the *system reciprocal* to $\mathbf{a}$, $\mathbf{b}$, and $\mathbf{c}$. They have peculiar and interesting properties which the student will find fully demonstrated in another work.*

It will be noticed that only two kinds of products of vectors have been defined, *i.e.*, the scalar product and the vector product. One should carefully remember as well that the scalar product and the vector product have been defined in terms of two simple vectors, but that instead of simple vectors any expression which is itself a vector may be used in place of the simple vectors to form these products. If this is carefully kept in mind it will make clear that in vector analysis certain combinations of symbols are meaningless.

For example, $\mathbf{a}\times\mathbf{b}$ being a vector, it may be used in conjunction with another simple vector $\mathbf{e}$, or another vector product $\mathbf{c}\times\mathbf{d}$, to form new scalar products and vector products, such as

$$(\mathbf{a}\times\mathbf{b})\cdot\mathbf{e} \quad \text{and} \quad (\mathbf{a}\times\mathbf{b})\cdot(\mathbf{c}\times\mathbf{d}),$$

$$(\mathbf{a}\times\mathbf{b})\times\mathbf{e} \quad \text{and} \quad (\mathbf{a}\times\mathbf{b})\times(\mathbf{c}\times\mathbf{d}),$$

or even

$$[(\mathbf{a}\times\mathbf{b})\times(\mathbf{c}\times\mathbf{d})]\times[(\mathbf{e}\times\mathbf{b})\times(\mathbf{g}\times\mathbf{h})], \text{ etc.},$$

which are all legitimate expressions. But, on the other hand, neither $\mathbf{a}\cdot\mathbf{b}$, nor $(\mathbf{a}\times\mathbf{b})\cdot(\mathbf{c}\times\mathbf{d})$, nor $(\mathbf{a}\times\mathbf{b})\cdot\mathbf{e}$ may be used *again* to form either scalar or vector products because they are merely scalars.

* Gibbs-Wilson, Vector Analysis, pp. 82–92.

Equations of Plane, Line, and Sphere.

32. The Plane Perpendicular to a and Passing through the Terminus of b. Let **r** be the radius vector to any point in it. The projection of **r** upon **a** is evidently constant and equal to the projection of **b** upon **a** as long as the terminus of **r** is in the plane; this condition is expressed by the equation

$$\mathbf{a}\cdot\mathbf{r} = \text{constant} = \mathbf{a}\cdot\mathbf{b}, \tag{64}$$

or

$$\mathbf{a}\cdot(\mathbf{r} - \mathbf{b}) = 0,$$

which is, therefore, the equation of the plane. It also states that $(\mathbf{r} - \mathbf{b})$ is perpendicular to **a** and hence parallel to the

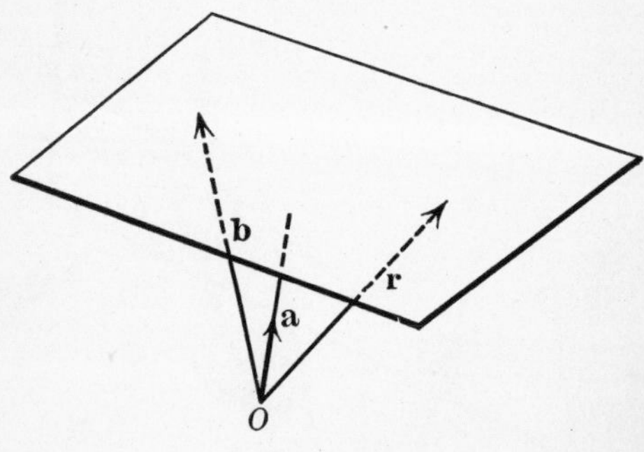

FIG. 33.

plane which is an evident truth and could be used to derive the equation of the plane. If the origin is in the plane, $\mathbf{b} = 0$ and

$$\mathbf{a}\cdot\mathbf{r} = 0 \tag{65}$$

is the equation, which is otherwise evident, as **r** is then always perpendicular to **a**. If the equation of a plane is desired, the plane being parallel to two given vectors **c** and **d** and passing through the terminus of **b**, simply remember that $\mathbf{c}\times\mathbf{d}$ is a vector perpendicular to the plane, and putting $\mathbf{c}\times\mathbf{d}$ in place of **a** above, its equation is

$$(\mathbf{c}\times\mathbf{d})\cdot(\mathbf{r} - \mathbf{b}) = 0. \tag{66}$$

If the equation of a plane passing through the ends of three given vectors a, b, and c is desired, remember that the vectors $(\mathbf{r} - \mathbf{a})$, $(\mathbf{a} - \mathbf{b})$, and $(\mathbf{b} - \mathbf{c})$ lie in the same plane and express this fact by (50), giving

$$(\mathbf{r} - \mathbf{a})\cdot(\mathbf{a} - \mathbf{b})\times(\mathbf{b} - \mathbf{c}) = 0,$$

or expanding

$$\left.\begin{aligned} &\mathbf{r}\cdot(\mathbf{a}\times\mathbf{b} + \mathbf{b}\times\mathbf{c} + \mathbf{c}\times\mathbf{a}) = \mathbf{a}\cdot(\mathbf{a}\times\mathbf{b} + \mathbf{b}\times\mathbf{c} + \mathbf{c}\times\mathbf{a}) \\ \text{or}\quad &\boldsymbol{\phi}\cdot(\mathbf{r} - \mathbf{a}) = 0, \text{ where } \boldsymbol{\phi} \equiv (\mathbf{a}\times\mathbf{b} + \mathbf{b}\times\mathbf{c} + \mathbf{c}\times\mathbf{a}). \end{aligned}\right\} \quad (67)$$

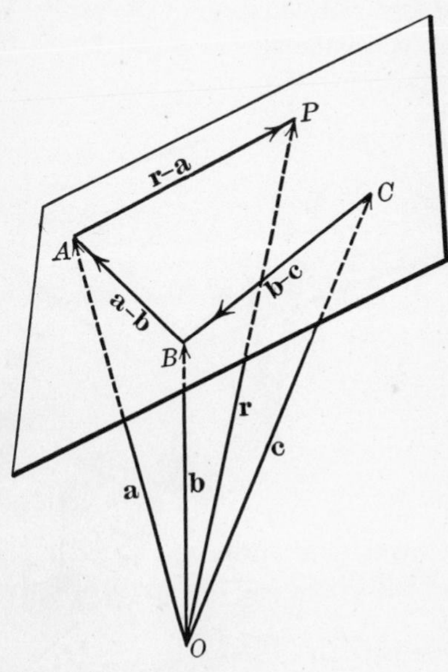

FIG. 34.

Comparing this last equation with (64), we see that $\boldsymbol{\phi}$ is a vector perpendicular to the plane.

To find the vector-perpendicular p from the origin to the plane. Referring to the plane in Fig. 33, the equation of which is

$$\mathbf{a}\cdot(\mathbf{r} - \mathbf{b}) = 0,$$

let $\mathbf{r}$ become perpendicular to the plane and hence some multiple of $\mathbf{a}$, say $x\,\mathbf{a}$, then

$$\mathbf{a}\cdot(x\,\mathbf{a} - \mathbf{b}) = 0,$$

or $$x\,\mathbf{a}^2 = \mathbf{a}\cdot\mathbf{b}; \qquad \text{then} \quad x = \frac{\mathbf{a}\cdot\mathbf{b}}{\mathbf{a}^2}$$

and $$\mathbf{p} = x\,\mathbf{a} = \mathbf{a}\,\frac{\mathbf{a}\cdot\mathbf{b}}{\mathbf{a}^2} = \frac{1}{\mathbf{a}}\,\mathbf{a}\cdot\mathbf{b} = \mathbf{a}^{-1}\,\mathbf{a}\cdot\mathbf{b}. \qquad (68a)$$

This result is also evident on inspection. For $\mathbf{a}\cdot\mathbf{b}$ is the projection of $\mathbf{b}$ on $\mathbf{a}$ multiplied by the magnitude of $\mathbf{a}$; hence, to obtain the value of the projection we must divide by the magnitude of $\mathbf{a}$, so that directly

$$\mathbf{p} = \frac{\mathbf{a}\cdot\mathbf{b}}{\mathbf{a}} = \mathbf{a}^{-1}\,\mathbf{a}\cdot\mathbf{b}.$$

In equation (67) the perpendicular $\mathbf{p}$ is then

$$\mathbf{p} = \phi^{-1}\,\phi\cdot\mathbf{a}. \qquad (68b)$$

33. The equation of a straight line through the end of b parallel to a is, since $\mathbf{a}$ and $(\mathbf{r} - \mathbf{b})$ are always parallel,

$$\mathbf{a}\times(\mathbf{r} - \mathbf{b}) = 0. \qquad (69)$$

This is compatible with the equation derived previously,

$$\mathbf{r} = \mathbf{b} + x\,\mathbf{a},$$

for applying $\mathbf{a}\times$ to it, we obtain 69 in the form

$$\mathbf{a}\times\mathbf{r} = \mathbf{a}\times\mathbf{b}.$$

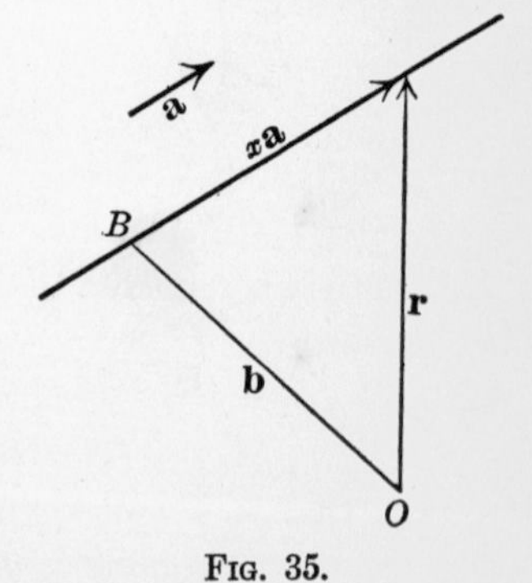

Fig. 35.

Again, the equation of a line perpendicular to $\mathbf{c}$ and $\mathbf{d}$ and passing through the end of $\mathbf{b}$ is, because $\mathbf{c}\times\mathbf{d}$ is parallel to $\mathbf{a}$ in the above equation,

$$(\mathbf{c}\times\mathbf{d})\times(\mathbf{r} - \mathbf{b}) = 0. \qquad (70)$$

34. Equations of the Circle and the Sphere. In a circle or sphere, with origin at the center, the length of the vector to the surface from the origin is constant and equal to the radius. Hence

$$r = a, \text{ (not } \mathbf{r} = \mathbf{a})$$

or

$$\mathbf{r}^2 = \mathbf{a}^2, \tag{71}$$

is the equation of the circle or of the sphere according as two or three dimensional space is considered.

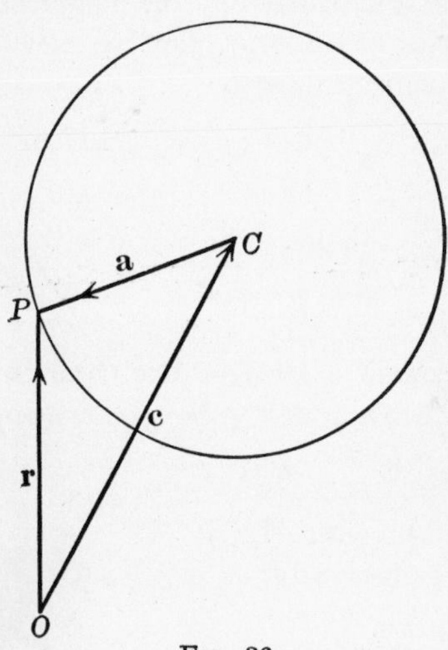

FIG. 36.

If the origin is removed to O at a distance $-\mathbf{c}$ from O, then (71) becomes

$$(\mathbf{r} - \mathbf{c})^2 = \mathbf{a}^2,$$

or

$$\mathbf{r}^2 - 2\,\mathbf{r}\cdot\mathbf{c} = \mathbf{a}^2 - \mathbf{c}^2 = \text{const.} \tag{72}$$

If $\mathbf{c} = \mathbf{a}$, that is, if the origin is on the circumference of the circle, the equation reduces to

$$\mathbf{r}^2 - 2\,\mathbf{r}\cdot\mathbf{a} = 0. \tag{73}$$

This, in *polar* coördinates, is nothing more than

$$r = 2\,a \cos \theta,$$

where θ is the angle between $\mathbf{r}$ and any predetermined radius, from which θ is measured.

In *rectangular* coördinates (73) considered as the equation of a sphere is written immediately

$$x^2 + y^2 + z^2 = 2\,(xa_1 + ya_2 + za_3),$$

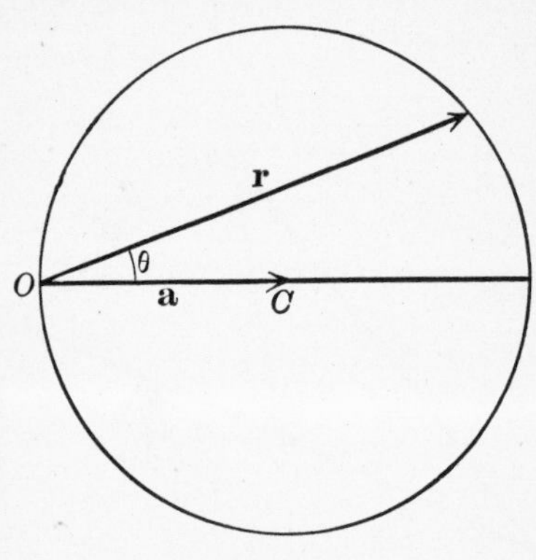

Fig. 37.

where a_1, a_2, a_3 are the projections of any chosen radius along the three axes. If the plane of yz be taken perpendicular to this radius, a_2 and a_3 are zero, so that

$$x^2 + y^2 + z^2 = 2\,a_1 x.$$

If we drop the z-coördinate the equation reduces to that of a circle tangent at the origin to the y axis and with its center on the x axis.

The equation with origin at center may be put in the form

$$\mathbf{r}^2 - \mathbf{a}^2 = 0,$$

or

$$(\mathbf{r} - \mathbf{a})\cdot(\mathbf{r} + \mathbf{a}) = 0,$$

which says, see Fig. 38, that the two lines AD and DB are always at right angles, a familiar result. Such illustrations may be multiplied indefinitely and show the ease with which equations may be written down to fit almost any conditions. When translated into their Cartesian equivalents they give familiar forms.

Such books as Tait, "Quaternions," Kelland and Tait, "Introduction to Quaternions," may be consulted with advantage at this point by the student. In these books the whole treatment of the line, plane, circle, sphere, and conic sections, with few exceptions, is one of vector analysis pure and

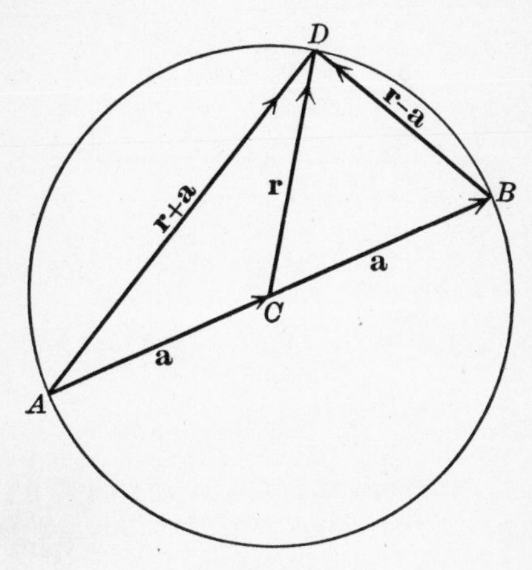

FIG. 38.

simple. The occurrence of a quaternion is a very rare event. The only difference to be noted particularly in reading these works is that the scalar product has the opposite sign to ours, and that

$$\mathbf{a}\cdot\mathbf{b} \text{ is written } S\,\mathbf{ab}$$

and

$$\mathbf{a}\times\mathbf{b} \text{ is written } V\,\mathbf{ab}.$$

34a. Resolution of a System of Forces Acting on a Rigid Body. Consider any point O as origin. This point may be anywhere, even outside of the body. The system of forces

$\mathbf{F}_1$, $\mathbf{F}_2$, $\mathbf{F}_3$, etc., acting on the body is equivalent to a single force $\mathbf{R}$ at O, where

$$\mathbf{R} = \sum \mathbf{F}$$

and a couple whose strength is

$$\mathbf{C} = \sum \mathbf{a} \times \mathbf{F},$$

where $\mathbf{a}_1$, $\mathbf{a}_2 \ldots$ are the vectors to the forces $\mathbf{F}_1$, $\mathbf{F}_2 \ldots$ from O.

To prove this, consider one of the forces $\mathbf{F}$ acting at P; we may introduce the zero system $+\mathbf{F}-\mathbf{F}$ at O without alter-

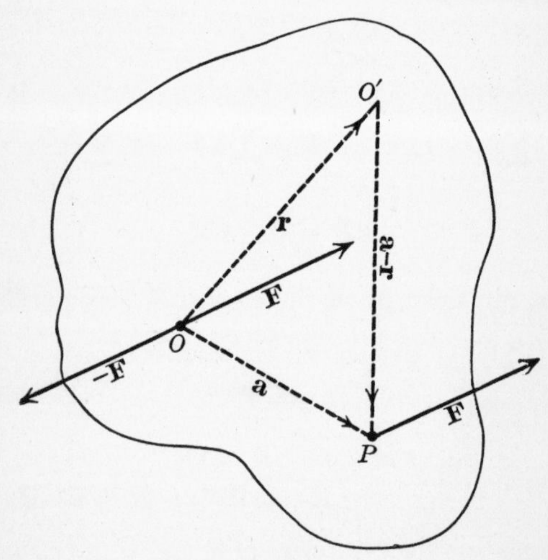

FIG. 39.

ing in any way the effect of $\mathbf{F}$ on the body. By combining $-\mathbf{F}$ with the $\mathbf{F}$ at P we get a couple of strength $\mathbf{a} \times \mathbf{F}$, leaving a force $\mathbf{F}$ acting at O. We may do the same for each of the forces $\mathbf{F}_1$, $\mathbf{F}_2$, etc., so that finally we have

(*a*) all the forces $\mathbf{F}_1, \ldots \mathbf{F}_2$ acting at O

and (*b*) an equal number of couples $\mathbf{a}_1 \times \mathbf{F}_1$, $\mathbf{a}_2 \times \mathbf{F}_2 \ldots$

Combine all the forces into a resultant force, $\mathbf{R} = \sum \mathbf{F}$ and all the couples in a resultant couple $\mathbf{C} = \sum \mathbf{a} \times \mathbf{F}$, which proves the theorem. Such a resolution may be made for any point whatever.

Central Axis. Couple a Minimum for Points on this Axis. In general the axis of the couple $\mathbf{C}$ derived above is not parallel to the resultant force $\mathbf{R}$.

Let us inquire if there are any points for which an analogous resolution will give a couple whose axis is parallel to $\mathbf{R}$. Notice that whatever point O' is chosen $\mathbf{R}$ is the same vector.

Let O' be such a point and let $OO' = \mathbf{r}$, and for O' the couple would be, since $\overline{O'P} = \mathbf{a} - \mathbf{r}$,

$$\sum (\mathbf{a} - \mathbf{r}) \times \mathbf{F}.$$

Then the condition that this couple be parallel to $\mathbf{R}$ becomes

$$\sum (\mathbf{a} - \mathbf{r}) \times \mathbf{F} = x\,\mathbf{R} = \sum \mathbf{a} \times \mathbf{F} - \mathbf{r} \times \sum \mathbf{F},$$

so that

$$x\,\mathbf{R} = \mathbf{C} - \mathbf{r} \times \mathbf{R}.$$

To find x, multiply by $\cdot\mathbf{R}^{-1}$, because $\mathbf{R}^{-1}$ is parallel to $\mathbf{R}$, and the triple scalar product term vanishes.

$$x = \mathbf{C} \cdot \mathbf{R}^{-1}.$$

Hence

$$\begin{aligned} \mathbf{r} \times \mathbf{R} &= \mathbf{C} - \mathbf{R}\,(\mathbf{C} \cdot \mathbf{R}^{-1}) \\ &= \mathbf{R} \times (\mathbf{C} \times \mathbf{R}^{-1}) = (\mathbf{R}^{-1} \times \mathbf{C}) \times \mathbf{R}, \end{aligned}$$

a *linear* equation for $\mathbf{r}$ in terms of $\mathbf{R}$ and $\mathbf{C}$ (given vectors by last theorem). There is then a *line* of points for which this kind of resolution is possible. This line is called the Central Axis of the System, which is then reduced to a force $\mathbf{R}$ and a couple *about* $\mathbf{R}$ of a certain magnitude $= x\mathbf{R} = \mathbf{R}\,\mathbf{C} \cdot \mathbf{R}^{-1}$

$$= \mathbf{R}_1 (\mathbf{C} \cdot \mathbf{R}_1). \qquad (a)$$

Considering the equation of the central axis

$$\mathbf{r} \times \mathbf{R} = (\mathbf{R}^{-1} \times \mathbf{C}) \times \mathbf{R},$$

it is seen that one of the values of $\mathbf{r}$ is $(\mathrm{R}^{-1}\times\mathrm{C})$, and since this last vector is evidently perpendicular to R, it must be the line ON; so that the equation may also be written

$$\mathbf{r} = \mathrm{R}^{-1}\times\mathrm{C} + y\mathrm{R} = \overline{ON} + y\mathrm{R},$$

where $\mathrm{R}^{-1}\times\mathrm{C}$ is the normal vector from O to the central axis.

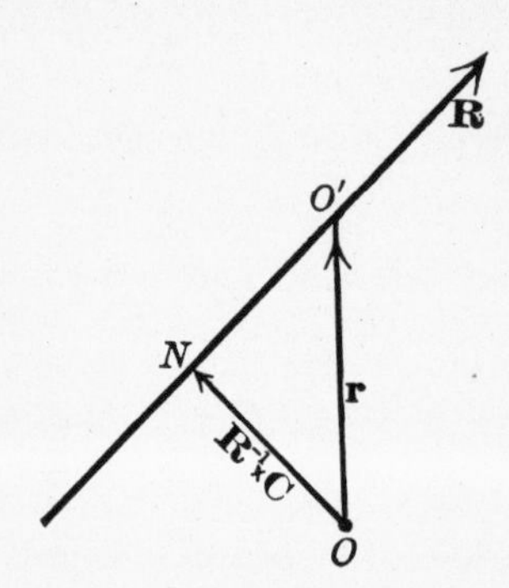

FIG. 39A.

By (a) we see that the couple about the central axis is the *component* of the couple at any other point along R, and hence is always *less* than for any other point; so that it is a minimum and equal to

$$\mathrm{C}\cdot\mathrm{R}_1 \text{ along R}.$$

It is, however, the same, *i.e.*, constant for *all* points on the central axis.

EXERCISES AND PROBLEMS.

1. Prove the following formulæ:

$$\mathbf{a}\times\{\mathbf{b}\times(\mathbf{c}\times\mathbf{d})\} = [\mathbf{acd}]\,\mathbf{b} - \mathbf{a}\cdot\mathbf{b}\ \mathbf{c}\times\mathbf{d}$$
$$= \mathbf{b}\cdot\mathbf{d}\ \mathbf{a}\times\mathbf{c} - \mathbf{b}\cdot\mathbf{c}\ \mathbf{a}\times\mathbf{d}$$

$$[\mathbf{a}\times\mathbf{b}\ \mathbf{c}\times\mathbf{d}\ \mathbf{e}\times\mathbf{f}] = [\mathbf{abd}]\,[\mathbf{cef}] - [\mathbf{abc}]\,[\mathbf{def}]$$
$$= [\mathbf{abe}]\,[\mathbf{fcd}] - [\mathbf{abf}]\,[\mathbf{ecd}]$$
$$= [\mathbf{cda}]\,[\mathbf{bef}] - [\mathbf{cdb}]\,[\mathbf{aef}].$$

2. Prove

$$[\mathbf{a}\times\mathbf{b}\ \mathbf{b}\times\mathbf{c}\ \mathbf{c}\times\mathbf{a}] = [\mathbf{abc}]^2$$

and interpret the result by determinants.

3. Show that

$$[\mathbf{pqr}]\,(\mathbf{a}\times\mathbf{b}) = \begin{vmatrix} \mathbf{p}\cdot\mathbf{a} & \mathbf{p}\cdot\mathbf{b} & \mathbf{p} \\ \mathbf{q}\cdot\mathbf{a} & \mathbf{q}\cdot\mathbf{b} & \mathbf{q} \\ \mathbf{r}\cdot\mathbf{a} & \mathbf{r}\cdot\mathbf{b} & \mathbf{r} \end{vmatrix}.$$

4. Show that

$$\mathbf{a}\times(\mathbf{b}\times\mathbf{c}) + \mathbf{b}\times(\mathbf{c}\times\mathbf{a}) + \mathbf{c}\times(\mathbf{a}\times\mathbf{b}) = 0.$$

5. Show that

$$[\mathbf{a}\times\mathbf{p}\ \mathbf{b}\times\mathbf{q}\ \mathbf{c}\times\mathbf{r}] + [\mathbf{a}\times\mathbf{q}\ \mathbf{b}\times\mathbf{r}\ \mathbf{c}\times\mathbf{p}] + [\mathbf{a}\times\mathbf{r}\ \mathbf{b}\times\mathbf{p}\ \mathbf{c}\times\mathbf{q}] = 0.$$

6. Prove that

$$(\mathbf{a}\times\mathbf{b})\cdot(\mathbf{c}\times\mathbf{d}) = (\mathbf{a}\cdot\mathbf{c})(\mathbf{b}\cdot\mathbf{d}) - (\mathbf{a}\cdot\mathbf{d})(\mathbf{b}\cdot\mathbf{c})$$

and that

$$\begin{aligned}(\mathbf{a}\times\mathbf{b})\times(\mathbf{c}\times\mathbf{d}) &= \mathbf{b}\,[\mathbf{acd}] - \mathbf{a}\,[\mathbf{bcd}] \\ &= \mathbf{c}\,[\mathbf{abd}] - \mathbf{d}\,[\mathbf{abc}].\end{aligned}$$

7. Deduce the fundamental formulæ of spherical trigonometry from the equations

$$\begin{aligned}(\mathbf{a}\times\mathbf{b})\cdot(\mathbf{c}\times\mathbf{d}) &= \mathbf{a}\cdot\mathbf{c}\ \mathbf{b}\cdot\mathbf{d} - \mathbf{a}\cdot\mathbf{d}\ \mathbf{b}\cdot\mathbf{c}, \\ (\mathbf{a}\times\mathbf{b})\times(\mathbf{c}\times\mathbf{d}) &= [\mathbf{acd}]\,\mathbf{b} - [\mathbf{bcd}]\,\mathbf{a} \\ &= [\mathbf{abd}]\,\mathbf{c} - [\mathbf{abc}]\,\mathbf{d}.\end{aligned}$$

Make the vectors unit vectors and take an origin at center of sphere of unit radius, thus making all the vectors terminate upon the surface of the sphere.

8. Show that the components of **b** parallel and perpendicular to **a** are respectively

$$\mathbf{b}' = \mathbf{a}\,\frac{\mathbf{a}\cdot\mathbf{b}}{\mathbf{a}\cdot\mathbf{a}} = \mathbf{a}_1\,\mathbf{a}_1\cdot\mathbf{b}$$

and

$$\mathbf{b}'' = -\,\frac{\mathbf{a}\times(\mathbf{a}\times\mathbf{b})}{\mathbf{a}\cdot\mathbf{a}} = -\,\mathbf{a}_1\times(\mathbf{a}_1\times\mathbf{b}).$$

9. The second vector **a** may be omitted from $\mathbf{a}\times(\mathbf{a}+\mathbf{b})$. May it be omitted in $\mathbf{a}^{-1}\times(\mathbf{a}+\mathbf{b})$ or in $\mathbf{a}\times(\mathbf{a}+\mathbf{b})^{-1}$?

10. The perpendiculars from the vertices of a triangle to the sides opposite meet in a point.

11. Find the point of intersection of a line and a plane and discuss the result.

12. The perpendicular bisectors of the sides of a triangle meet in a point.

13. Find an expression for the common perpendicular to two lines not lying in the same plane.

14. Determine the vector-perpendicular drawn from the origin to the plane determined by the three points $\mathbf{a}$, $\mathbf{b}$, $\mathbf{c}$.

15. Find the equation of a plane passing through a given point $\mathbf{c}$ and parallel to each of two given straight lines $\mathbf{b}'$ and $\mathbf{b}''$.

Ans. $(\mathbf{r} - \mathbf{c}) \cdot \mathbf{b}' \times \mathbf{b}'' = 0$.

16. Find the length of the common perpendicular to each of two given straight lines parallel to $\mathbf{b}_1$ and $\mathbf{b}_2$ and passing through $\mathbf{a}_1$ and $\mathbf{a}_2$ respectively, and show that it is

$$\mathbf{d} = y\,(\mathbf{b}_1 \times \mathbf{b}_2),$$

where

$$y = \frac{(\mathbf{a}_1 - \mathbf{a}_2) \cdot (\mathbf{b}_1 \times \mathbf{b}_2)}{(\mathbf{b}_1 \times \mathbf{b}_2)^2}.$$

17. Find the equation of the line of intersection of two planes.

18. Deduce the Cartesian equation for the volume of the tetrahedron whose vertices are

$$\mathbf{a},\ \mathbf{b},\ \mathbf{c},\ \mathbf{d},$$

where

$$\mathbf{a} = a_1\mathbf{i} + a_2\mathbf{j} + a_3\mathbf{k}, \text{ etc.}$$

19. Deduce the Cartesian equation for the area of the triangle whose vertices are

$$\mathbf{a} = a_1\mathbf{i} + a_2\mathbf{j} + a_3\mathbf{k},$$
$$\mathbf{b} = b_1\mathbf{i} + b_2\mathbf{j} + b_3\mathbf{k},$$
$$\mathbf{c} = c_1\mathbf{i} + c_2\mathbf{j} + c_3\mathbf{k}.$$

20. Find by translating into Cartesian notation that the volume of a pyramid, of which the vertex is a given point (xyz) and the base a triangle formed by joining three given points aoo, obo, ooc in the rectangular coördinate axes, is

$$V = \tfrac{1}{6}\,abc\left(\frac{x}{a} + \frac{y}{b} + \frac{z}{c} - 1\right).$$

21. Show how to determine the directions of two vectors of given magnitude so that their resultant shall be of given magnitude and direction. When is this impossible?

22. The moment of the force $\overline{AB}$ about the line $\overline{CD}$ is six times the volume of the tetrahedron $ABCD$ divided by the number of units of length in CD.

Find vector moment at C, and then the component of this *along* CD.

23. The laws of refraction of light from a medium of index μ into one of index μ' are comprised in the relation

$$\mu\, \mathbf{n}{\times}\mathbf{a} = \mu'\, \mathbf{n}{\times}\mathbf{a}',$$

where $\mathbf{n}$, $\mathbf{a}$, and $\mathbf{a}'$ are unit vectors along the normal, the incident and refracted rays respectively.

24. Write out the equations of problems 1 to 8 inclusive in Cartesian notation.

25. If $\mathbf{r} = \mathbf{a} + \mathbf{b}\, x$ is the equation of a straight line, $\mathbf{b}$ being a unit vector, prove that the line through the origin perpendicular to it is

$$\mathbf{r} = y\ (\mathbf{a} - \mathbf{a}{\cdot}\mathbf{b}\ \mathbf{b})$$

and that its length is

$$\sqrt{\mathbf{a}^2 - (\mathbf{a}{\cdot}\mathbf{b})^2}.$$

26. The equation of the plane through the origin perpendicular to the vector $\mathbf{a}$ may be written in either of the forms

$$\mathbf{a}{\cdot}\mathbf{r} = 0 \qquad \mathbf{a}^{-1}{\cdot}\mathbf{r} = 0 \qquad (\mathbf{r} + \mathbf{a})_0 = (\mathbf{r} - \mathbf{a})_0.$$

27. Let $\mathbf{r} = \mathbf{b} + x\,\mathbf{d}$ be the equation of a straight line. For what values of x does it meet the sphere $\mathbf{r}^2 = \mathbf{c}^2$? By the theorem on the coefficients of an equation as related to the roots, derive the theorems for the product and sum of the intercepts respectively.

28. Derive the equation of the sphere (or circle) from the equation

$$(\mathbf{r} - \mathbf{c})_0 = a.$$

This equation states that the end of $\mathbf{r}$ must remain at a constant distance a from the end of $\mathbf{c}$, hence, etc. . . .

29. Prove that if the sum and difference of two forces are at right angles the two forces are equal.

30. Prove that if the lengths of the sum and difference respectively of two forces are equal, the two forces are at right angles.

CHAPTER IV.

DIFFERENTIATION OF VECTORS.

35. Two Ways in which a Vector may Vary. If a small vector $d\mathbf{a}$ be added to the vector $\mathbf{a}$, the result in general will be a new vector differing by a small amount from $\mathbf{a}$ not only in *length* but also in *direction*.

If the small vector which is added be perpendicular to $\mathbf{a}$, then the *length* of $\mathbf{a}$ will remain unchanged.

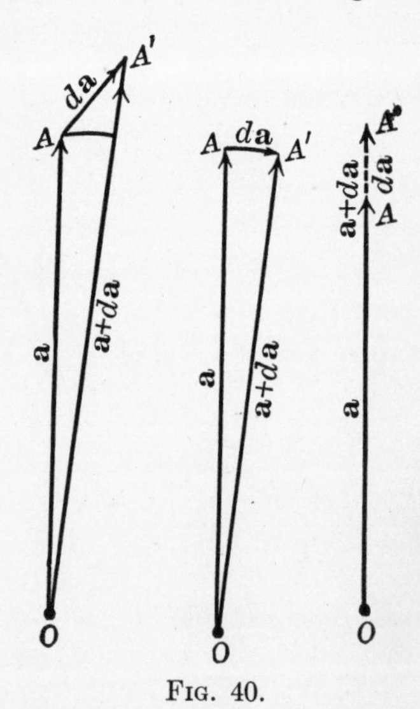

Fig. 40.

If the small vector which is added be parallel to $\mathbf{a}$, then its *direction* will remain unchanged.

These three cases are shown in Fig. 40.

Differentiation with Respect to Scalar Variables. Let the vector $\mathbf{a}\,(t)$ be a function of a scalar variable t. In other words, let its length and direction be known and determinate as soon as a value of t is given.

Let $\overline{OA_1}$ be the value of $\mathbf{a}(t)$ when $t = t_1$, and let $\overline{OA_2}$ be the new value of $\mathbf{a}(t)$ when $t = t_2$. Then the change in $\mathbf{a}(t)$ due to the change in t of $t_2 - t_1$ is

$$\overrightarrow{A_1A_2} = \mathbf{a}(t_2) - \mathbf{a}(t_1).$$

Dividing this equation by $t_2 - t_1$ in order to find the *rate* of change and making $t_2 - t_1$ infinitely small, we define

$$\frac{d\mathbf{a}}{dt} = \lim_{t_2 - t_1 \doteq 0} \frac{\mathbf{a}(t_2) - \mathbf{a}(t_1)}{t_2 - t_1}. \quad (74)$$

If $t_2 - t_1 = h$, where h is some small scalar, this may be written in the more familiar form,

$$\frac{d\mathbf{a}}{dt} = \lim_{h \doteq 0} \frac{\mathbf{a}(t+h) - \mathbf{a}(t)}{h}. \quad (75)$$

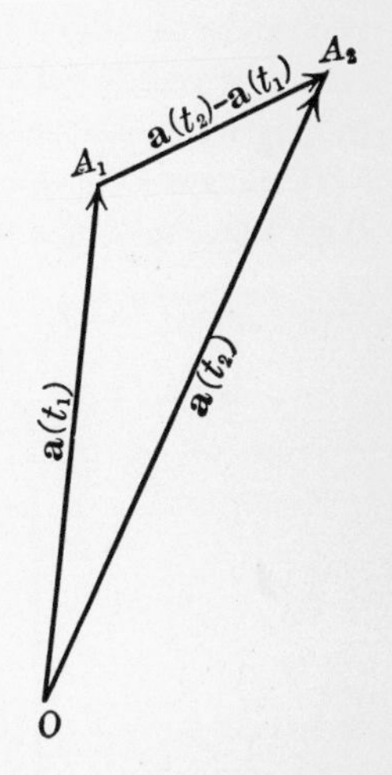

FIG. 41.

Evidently the rate change of the vector $\mathbf{a}$ with respect to t is made up vectorially of the three rates of change of its components

$\frac{da_1}{dt}$ along $\mathbf{i}$, $\frac{da_2}{dt}$ along $\mathbf{j}$, and $\frac{da_3}{dt}$ along $\mathbf{k}$,

or
$$\frac{d\mathbf{a}}{dt} = \frac{da_1}{dt}\mathbf{i} + \frac{da_2}{dt}\mathbf{j} + \frac{da_3}{dt}\mathbf{k}. \quad (76)$$

Precisely the same reasoning holds for the rate of change with respect to t of the vector representing $\frac{d\mathbf{a}}{dt}$,

$$\text{written} \quad \frac{d}{dt}\left(\frac{d\mathbf{a}}{dt}\right) \quad \text{or} \quad \frac{d^2\mathbf{a}}{dt^2}.$$

Similarly for the higher derivatives we may then write

$$\frac{d^n\mathbf{a}}{dt^n} = \frac{d^na_1}{dt^n}\mathbf{i} + \frac{d^na_2}{dt^n}\mathbf{j} + \frac{d^na_3}{dt^n}\mathbf{k}. \quad (77)$$

The vectors **i j k** are to be considered constant in length (being unit vectors) and in direction (being along fixed axes) in all of these differentiations.

If $\frac{d}{dt}$ be denoted by p, then

$$\frac{d\mathbf{a}}{dt} \equiv p\,\mathbf{a} = p\,(a_1\mathbf{i} + a_2\mathbf{j} + a_3\mathbf{k}) = pa_1\mathbf{i} + pa_2\mathbf{j} + pa_3\mathbf{k}$$

and (78)

$$\frac{d^n\mathbf{a}}{dt^n} \equiv p^n\mathbf{a} = p^n(a_1\mathbf{i} + a_2\mathbf{j} + a_3\mathbf{k}) = p^n a_1\mathbf{i} + p^n a_2\mathbf{j} + p^n a_3\mathbf{k}.$$

It will be noticed that the operator $p \equiv \frac{d}{dt}$ acts like a scalar multiplier.

36. Differentiation of Scalar and Vector Products. The differential of **a·b,** for instance, is defined just as in scalar calculus as

$$(\mathbf{a} + d\mathbf{a})\cdot(\mathbf{b} + d\mathbf{b}) - \mathbf{a}\cdot\mathbf{b} = d\,(\mathbf{a}\cdot\mathbf{b}).$$

Expanding and neglecting small quantities of the second order there remains

$$d\,(\mathbf{a}\cdot\mathbf{b}) = \mathbf{a}\cdot\mathbf{b} + d\mathbf{a}\cdot\mathbf{b} + \mathbf{a}\cdot d\mathbf{b} - \mathbf{a}\cdot\mathbf{b}$$
$$= d\mathbf{a}\cdot\mathbf{b} + \mathbf{a}\cdot d\mathbf{b};$$

hence, dividing through by dt,

$$\frac{d}{dt}\,(\mathbf{a}\cdot\mathbf{b}) = \frac{d\mathbf{a}}{dt}\cdot\mathbf{b} + \mathbf{a}\cdot\frac{d\mathbf{b}}{dt},$$

or $$p\,(\mathbf{a}\cdot\mathbf{b}) = p\,\mathbf{a}\cdot\mathbf{b} + \mathbf{a}\cdot p\,\mathbf{b} = \mathbf{b}\cdot p\,\mathbf{a} + p\,\mathbf{b}\cdot\mathbf{a}. \quad (79)$$

In a similar manner

$$\frac{d}{dt}\,(\mathbf{a}\times\mathbf{b}) = \frac{d\mathbf{a}}{dt}\times\mathbf{b} + \mathbf{a}\times\frac{d\mathbf{b}}{dt},$$

or $$p\,(\mathbf{a}\times\mathbf{b}) = p\,\mathbf{a}\times\mathbf{b} + \mathbf{a}\times p\,\mathbf{b}. \quad (80)$$

The differentiations then take place very much as they do in scalar calculus, but with this important difference, that the order of the factors must remain unchanged in all expressions where a change in order of the vectors, as previously explained, would not be allowable. For example, in (79) the order is immaterial, while in (80) it is essential.

The formulæ

$$\frac{d}{dt}[\mathbf{a}\cdot(\mathbf{b}\times\mathbf{c})] = \frac{d\mathbf{a}}{dt}\cdot(\mathbf{b}\times\mathbf{c}) + \mathbf{a}\cdot\left(\frac{d\mathbf{b}}{dt}\times\mathbf{c}\right) + \mathbf{a}\cdot\left(\mathbf{b}\times\frac{d\mathbf{c}}{dt}\right)$$

or $$p[\mathbf{a}\cdot(\mathbf{b}\times\mathbf{c})] = p\,\mathbf{a}\cdot(\mathbf{b}\times\mathbf{c}) + \mathbf{a}\cdot(p\,\mathbf{b}\times\mathbf{c}) + \mathbf{a}\cdot(\mathbf{b}\times p\,\mathbf{c}) \qquad (81)$$

and $$\frac{d}{dt}[\mathbf{a}\times(\mathbf{b}\times\mathbf{c})] = \frac{d\mathbf{a}}{dt}\times(\mathbf{b}\times\mathbf{c}) + \mathbf{a}\times\left(\frac{d\mathbf{b}}{dt}\times\mathbf{c}\right) + \mathbf{a}\times\left(\mathbf{b}\times\frac{d\mathbf{c}}{dt}\right)$$

or $$p[\mathbf{a}\times(\mathbf{b}\times\mathbf{c})] = p\,\mathbf{a}\times(\mathbf{b}\times\mathbf{c}) + \mathbf{a}\times(p\,\mathbf{b}\times\mathbf{c}) + \mathbf{a}\times(\mathbf{b}\times p\,\mathbf{c})$$

are in the same way easily seen to be true results.

It is instructive to notice the manner in which the operator p operates in turn on each one of the factors.

If a vector is to remain constant in length, then

$$a = \text{const.}$$

or $$\mathbf{a}\cdot\mathbf{a} = \text{const.};$$

hence $$\mathbf{a}\cdot d\mathbf{a} = 0,$$

or $d\mathbf{a}$ is perpendicular to **a**, as is geometrically evident by considering

$$\mathbf{a}\cdot\mathbf{a} = a^2 = \text{const.}$$

as the equation of a sphere or circle.

37. Applications to Geometry. We shall obtain some interesting and useful results, as well as a clearer insight into the calculus of vectors, by the following applications to geometry.

Let a variable vector **r** be drawn from a fixed origin O. We shall assume that the terminus of **r** can be located as soon as a value of t, an independent scalar variable, is given. By a slight extension of mathematical nomenclature and symbols we shall express this result by writing

$$\mathbf{r} = \mathbf{f}(t), \qquad (82)$$

reading it as: $\mathbf{r}$ equals a *vector* function of t. To indicate the vector character of the function, $\mathbf{f}$ is printed in bold-faced type.

As t varies continuously, the terminus of $\mathbf{r}$ describes some curve or curves in space, depending upon whether $\mathbf{r}$ is a single-valued or multi-valued function of t.

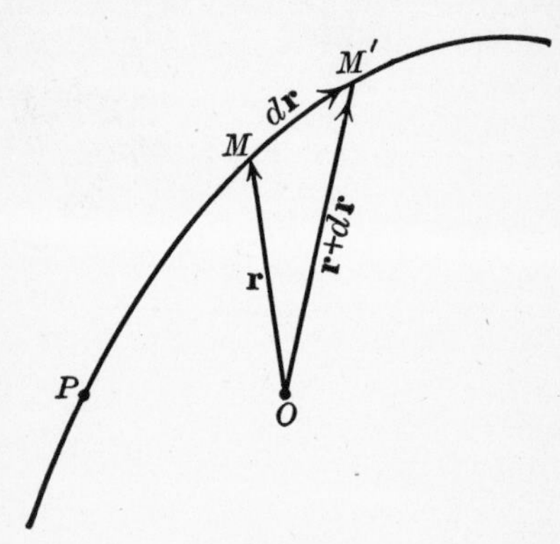

FIG. 42.

We assume in the following that the function $\mathbf{f}$ is a continuous and *single*-valued function of the independent scalar variable t.

Let $t = s$ be the distance along the curve

$$\mathbf{r} = \mathbf{f}\ (s)$$

from any point P on the curve. The increment $d\mathbf{r}$ is evidently a vector along the curve, and of length ds, hence $\frac{d\mathbf{r}}{ds}$ is a *unit* vector tangent to the curve at the point under consideration, M, when M' has approached indefinitely near to M. For convenience we shall write $\frac{d\mathbf{r}}{ds} \equiv \mathbf{t}$, where $\mathbf{t}$ is a unit vector along the tangent to the curve, or as we call it the *unit tangent*,

and
$$\frac{d\mathbf{r}}{ds} = \mathbf{t} = \frac{dx}{ds}\mathbf{i} + \frac{dy}{ds}\mathbf{j} + \frac{dz}{ds}\mathbf{k}$$
$$= t_1\mathbf{i} + t_2\mathbf{j} + t_3\mathbf{k}.$$

So that
$$t_1 = \frac{dx}{ds}, \quad t_2 = \frac{dy}{ds}, \quad t_3 = \frac{dz}{ds}$$

are the direction cosines of the tangent.

Tangent and Normal. The equation of the tangent line at $\mathbf{r}$ is then the equation of a line through the terminus of $\mathbf{r}$ and parallel to $\mathbf{t} = \frac{d\mathbf{r}}{ds}$, and by (69) is written

$$(\mathbf{r} - \boldsymbol{\xi}) \times \frac{d\mathbf{r}}{ds} = 0, \tag{83}$$

where $\boldsymbol{\xi}$ is the variable vector to this line from o.

Expanding this vector product by means of (40) we obtain the familiar Cartesian equations

$$\begin{vmatrix} \mathbf{i} & \mathbf{j} & \mathbf{k} \\ (x-\xi_1) & (y-\xi_2) & (z-\xi_3) \\ \frac{dx}{ds} & \frac{dy}{ds} & \frac{dz}{ds} \end{vmatrix} \equiv 0$$

or making the three components along $\mathbf{i}$, $\mathbf{j}$, and $\mathbf{k}$ equal to zero.

$$\frac{x-\xi_1}{\frac{dx}{ds}} = \frac{y-\xi_2}{\frac{dy}{ds}} = \frac{z-\xi_3}{\frac{dz}{ds}}. \tag{84}$$

The plane normal to the curve at $\mathbf{r}$ is, by (64),

$$(\mathbf{r} - \boldsymbol{\xi}) \cdot \frac{d\mathbf{r}}{ds} = 0, \tag{85}$$

or expanding in its Cartesian form,

$$(x-\xi_1)\frac{dx}{ds} + (y-\xi_2)\frac{dy}{ds} + (z-\xi_3)\frac{dz}{ds} = 0. \tag{86}$$

38. Curvature.* Consider three adjacent points on the curve, M_1, M_2, and M_3; the *unit* tangents through M_1 and M_2 and through M_2 and M_3 differ only in direction, hence the vector added to the first one to obtain the second one is at right angles to both and therefore measures the angle

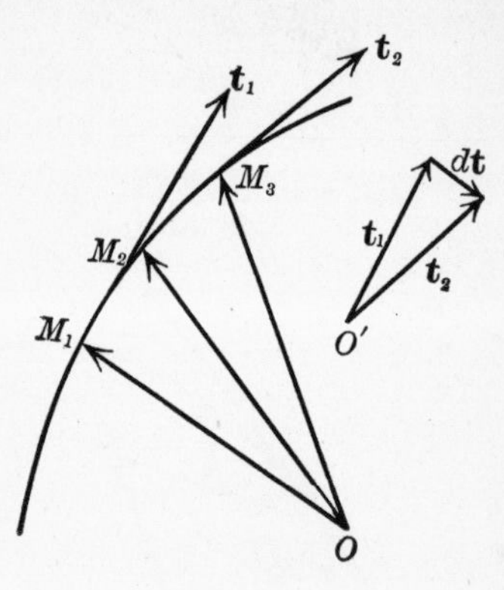

FIG. 43.

through which it is turned in going from M_1 to M_2. By definition the curvature is defined to be the magnitude of $\frac{d\mathbf{t}}{ds}$.

It is convenient to call the vector $\frac{d\mathbf{t}}{ds} = \frac{d^2\mathbf{r}}{ds^2}$ the vector-curvature, $\mathbf{c}$, as it has the same magnitude as $\frac{d\mathbf{t}}{ds}$ and, being normal to the tangent, points towards the center of curvature. The vector-curvature, being perpendicular to two consecutive tangents, lies in their plane. The radius of curvature $\boldsymbol{\rho}$ has a length inversely proportional to the magnitude of the curvature, but points in the same direction, and hence may be written

$$\boldsymbol{\rho} = \mathbf{c}^{-1} = \left(\frac{d^2\mathbf{r}}{ds^2}\right)^{-1}. \tag{87}$$

* See Appendix, p. 242, for other definitions.

Osculating Plane. The plane containing two consecutive tangents is called the osculating plane. If $\boldsymbol{\xi}$ be the vector to any point in this plane from O, since the three vectors $\mathbf{t}$, $\dfrac{d\mathbf{t}}{ds}$, and $(\mathbf{r} - \boldsymbol{\xi})$ lie in it, its equation may be written down at once by (50) as

$$(\mathbf{r} - \boldsymbol{\xi})\cdot\left(\mathbf{t}\times\frac{d\mathbf{t}}{ds}\right) = 0,$$

or

$$(\mathbf{r} - \boldsymbol{\xi})\cdot\left(\frac{d\mathbf{r}}{ds}\times\frac{d^2\mathbf{r}}{ds^2}\right) = 0. \qquad (88)$$

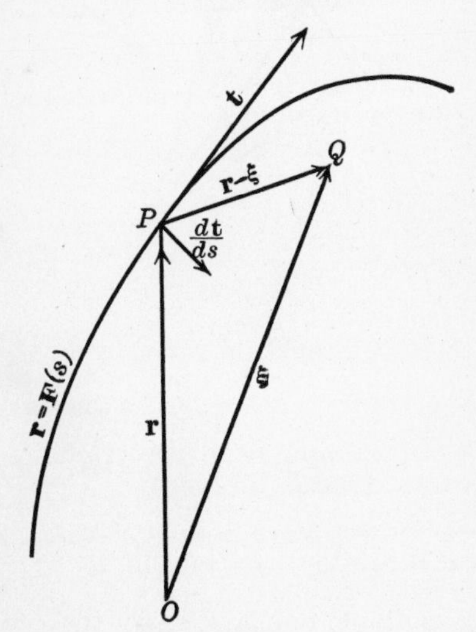

FIG. 44.

This by (49) may be written in the familiar form of a determinant

$$\begin{vmatrix} \xi_1 - x & \dfrac{dx}{ds} & \dfrac{d^2x}{ds^2} \\ \xi_2 - y & \dfrac{dy}{ds} & \dfrac{d^2y}{ds^2} \\ \xi_3 - z & \dfrac{dz}{ds} & \dfrac{d^2z}{ds^2} \end{vmatrix} = 0. \qquad (89)$$

Tortuosity. A twisted curve in space twists in two distinct ways. Any small portion of the curve lies in its osculating plane at that point, and this small portion of the curve has a curvature as described above. As we go along the curve, however, the osculating plane turns through a certain angle; the limit of the ratio of the angle turned through by the osculating plane to the arc traversed to produce that change is called the tortuosity.

Hence if $\mathbf{n}$ be a unit normal to the osculating plane,

$$\mathbf{T} = \frac{d\mathbf{n}}{ds},$$

where ds is the magnitude of the arc.

Geodetic Lines on a Surface. The differential equation to a geodetic line on a surface may be obtained in the following simple manner from the definition:

A geodetic line is a curve on a surface, the osculating plane of the curve being everywhere normal to the surface.

It is the curve a stretched string would lie along if the surface were a perfectly smooth one, the reaction of the surface to the pressure of the string being everywhere along the normal to the surface where it is in contact with the string.

Let $\mathbf{t}$ be the unit tangent to the geodetic, let $\mathbf{n}$ be the unit normal to the curve, and let $\mathbf{m} = \mathbf{n}\times\mathbf{t}$ be a unit vector lying therefore in the surface normal to $\mathbf{n}$ and to $\mathbf{t}$.

The osculating plane is determined by $\mathbf{t}$ and $d\mathbf{t}$ which lie in it by definition. If the curve is a geodetic, the normal to this plane $\mathbf{t}\times d\mathbf{t}$ lies in the surface, and is hence perpendicular to $\mathbf{n}$.

Expressing this fact,

$$\mathbf{n}\cdot(\mathbf{t}\times d\mathbf{t}) = 0.$$

Since $\mathbf{t}$ lies along $d\mathbf{r}$ (§ 37) and $d\mathbf{t}$ lies along $d^2\mathbf{r}$ (87), this equation becomes

$$\mathbf{n}\cdot(d\mathbf{r}\times d^2\mathbf{r}) = 0, \qquad (89a)$$

which is the differential equation to the geodetic.

If the surface is of the form

$$V(\mathbf{r}) = \text{const.},$$

then ∇V is a vector normal to it. See (106). But ∇V has direction cosines proportional to

$$\frac{\partial V}{\partial x}, \quad \frac{\partial V}{\partial y}, \quad \frac{\partial V}{\partial z},$$

and

$$\mathbf{n} \quad \text{is along} \quad \mathbf{i}\,\frac{\partial V}{\delta x} + \mathbf{j}\,\frac{\partial V}{\partial y} + \mathbf{k}\,\frac{\partial V}{\partial z}.$$

Hence (89a) becomes, by (49),

$$\begin{vmatrix} \frac{\partial V}{\partial x} & \frac{\partial V}{\partial y} & \frac{\partial V}{\partial z} \\ dx & dy & dz \\ d^2x & d^2y & d^2z \end{vmatrix} = 0.$$

39. Equations of Surfaces. Curvilinear Coördinates. The equation

$$\mathbf{r} = \mathbf{f}(u, v), \tag{90}$$

where u and v are two independent scalar variables, represents a surface.

If a particular value u_1 be given to u while v is unrestricted,

$$\mathbf{r} = \mathbf{f}(u_1, v)$$

being of the form (82), is some curve lying wholly on the surface. If a particular value v_1 be given to v while u is unrestricted,

$$\mathbf{r} = \mathbf{f}(u, v_1)$$

is some other curve lying wholly on the surface. These two curves intersect at the point or points $\mathbf{r}$ determined by the equation

$$\mathbf{r}_1 = \mathbf{f}(u_1, v_1).$$

We may then determine a point on the surface by giving particular values u_1 and v_1, say, to u and v. This point will be found at the intersection of the two curves

$$\mathbf{r} = \mathbf{f}\,(u_1, v) \text{ and } \mathbf{r} = \mathbf{f}\,(u, v_1).$$

Curvilinear coördinates is the name given to these variables u and v, such a series of curves divides the surface up into a network of curvilinear quadrilaterals, the angles of which may have any value. In the particular case that these curves cut each other always at right angles they

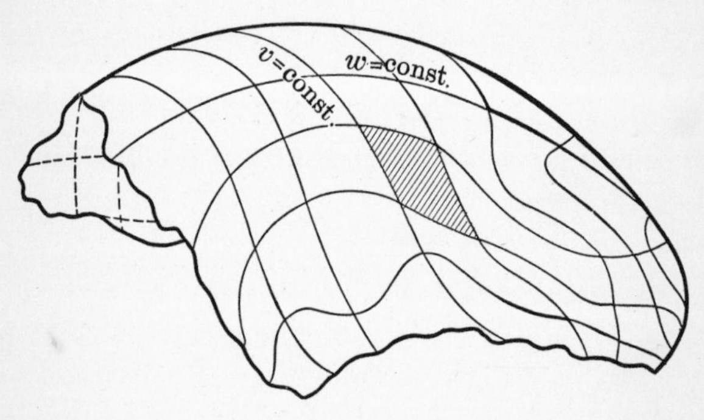

FIG. 45.

are said to form an **Orthogonal System** of curves. When the two systems of curves divide the surface up into infinitesimal *square elements* they are said to form an **Isothermal System.** Such systems are of the greatest importance in mathematical physics. The student should consult on this subject an excellent book by Fehr, "Applications de la Méthode Vectorielle à la Géométrie Infinitésimale" (Carré et Naud, Paris, 1899). His notation is different from ours, and is fully explained in his introduction, but his methods are quite similar.

40. Applications to the Kinematics of a Particle. Let the independent variable t now denote the time; then $\frac{d\mathbf{r}}{dt} = \mathbf{v}$ is the vector velocity along the curve $\mathbf{r} = \mathbf{f}\,(t)$. Notice that

$\frac{d\mathbf{r}}{dt}$ is no longer a unit vector, as here $dt \neq ds$, the element of arc, but the direction is still along the tangent. If $\mathbf{t}$ is the unit tangent to the curve, then

$$\mathbf{v} = v\,\mathbf{t},$$

where v, the magnitude of $\mathbf{v}$, is called the "speed."

The acceleration

$$\mathbf{a} = \frac{d\mathbf{v}}{dt} = \frac{d}{dt}(v\mathbf{t}) = \frac{dv}{dt}\mathbf{t} + v\frac{d\mathbf{t}}{dt}, \tag{91}$$

$\frac{dv}{dt}$ is the increase of speed *along* the curve. Speed is here used to denote the velocity irrespective of its direction.

And by (87)

$$\frac{d\mathbf{t}}{dt} = \frac{d\mathbf{t}}{ds}\frac{ds}{dt} = \mathbf{c}\,v,$$

$$\therefore \mathbf{a} = \frac{dv}{dt}\mathbf{t} + v^2\mathbf{c} = \frac{dv}{dt}\mathbf{t} + \frac{v^2}{\boldsymbol{\rho}}, \tag{92}$$

or the acceleration of a particle on a curve may be resolved into two components at right angles to each other, one $\frac{dv}{dt}$ increasing the linear speed along the curve, the other one $v^2\mathbf{c}$, or $\frac{v^2}{\boldsymbol{\rho}}$, where $\boldsymbol{\rho}$ is the vector radius of curvature and is directed towards the center of curvature, merely changing the direction of the motion.

Hodograph. The hodograph is a curve obtained in any given case of motion of a particle, by laying off from an arbitrary origin vectors equal to the velocities of the particle for all points of the path. The locus of the extremities of these vectors is the hodograph.

When a particle describes a curve, there is then a point related to it simultaneously describing the hodograph. This conception was introduced by Hamilton, and is an efficient aid to the study of curvilinear motion.

Evidently the hodograph itself may have a hodograph, and this perhaps another, depending upon the complexity of the motion, and so on.

The hodograph of a particle at rest is a point at the arbitrary origin.

The hodograph of uniform motion in a straight line is a point at the end of a vector of length equal to the velocity.

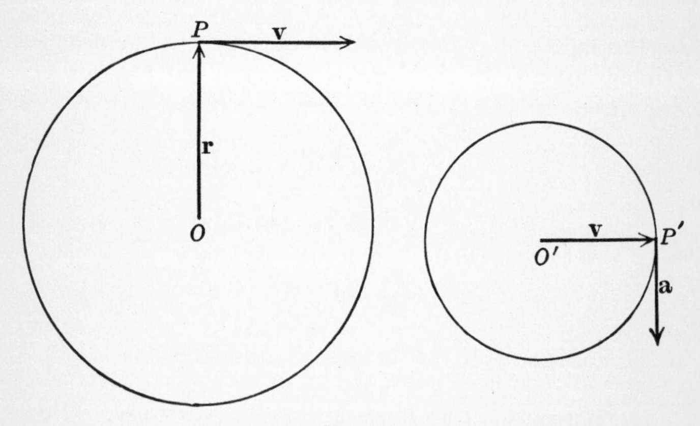

Fig. 46.

The hodograph of uniformly accelerated motion in a straight line is another straight line parallel to the first, described with *uniform* speed by the hodograph variable.

The hodograph of uniform motion in a circle is another circle, since the speed is constant, of radius equal to the speed. The vector velocity in the circle is always perpendicular to the radius vector in the original path. Evidently the points P and P' move around their origins with the same angular velocities. The velocity of P' in general is the rate of change of v, and hence is the acceleration of P.

Since the two circles must be described in the same times,

$$\frac{2\pi r}{v} = \frac{2\pi v}{a};$$

hence

$$a = \frac{v^2}{r}$$

a familiar result.

Equation of the Hodograph. If

$$\mathbf{r} = \mathbf{f}(t)$$

be the equation of the path described by a particle, containing not merely the form of the path but the law of its description as well, then

$$\frac{d\mathbf{r}}{dt} = \mathbf{f}'(t)$$

is the equation of the hodograph and the law of its description. Again

$$\frac{d^2\mathbf{r}}{dt^2} = \mathbf{f}''(t)$$

is the hodograph of the hodograph, and so on.

41. Integration with Respect to Scalar Variables. (Reconsult paragraphs 4 and 16.)

The inverse of differentiation offers merely the difficulties of scalar integration. The constants of integration, however, are constant vectors. As a simple example consider the motion of a particle under constant acceleration, under gravity for instance. The differential equation of the motion may be written

$$\frac{d^2\mathbf{r}}{dt^2} = \mathbf{a},$$

where **a** is a constant vector. Integrating once,

$$\frac{d\mathbf{r}}{dt} = \mathbf{a}\,t + \mathbf{v}_0,$$

where $\mathbf{v}_0$ is a constant *vector*, as it is a vector equation and $\mathbf{v}_0$ is determined by the value of $\frac{d\mathbf{r}}{dt}$ when $t = 0$. Integrating again we obtain

$$\mathbf{r} = \tfrac{1}{2}\,\mathbf{a}\,t^2 + \mathbf{v}_0 t + \mathbf{s}_0, \tag{93}$$

where again $\mathbf{s}_0$ is a second constant vector and whose value is that of $\mathbf{r}$ when $t = 0$. Equation (93) gives the value of $\mathbf{r}$ at any time t. The equation says that starting from the point $\mathbf{s}_0$ (*i.e.* at S), $\mathbf{r}$, the vector to any point of the path, may

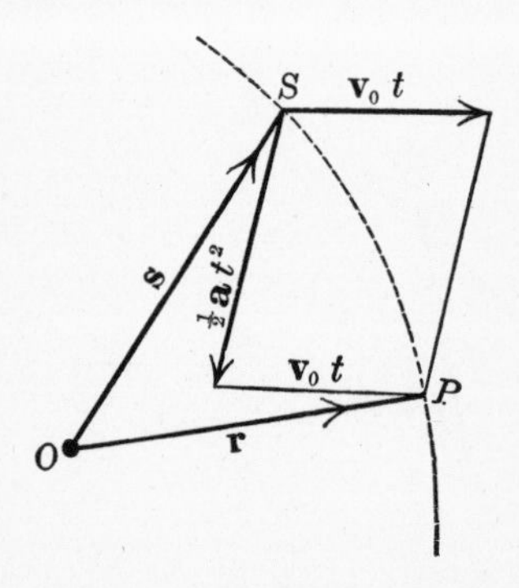

FIG. 46A.

be found by adding to $\mathbf{s}_0$ the vector sum of the two motions $\mathbf{v}_0 t$ and $\frac{1}{2}\,\mathbf{a}\,t^2$. The terminus of $\mathbf{r}$ evidently describes a parabola passing through $\mathbf{s}_0$, because the coördinates of any point on the curve referred to the oblique axes parallel to $\mathbf{v}_0$ and $\mathbf{a}$ are proportional to the first power and the second power of the same quantity t, respectively.

Orbit of a Planet. Central Acceleration. As another example, consider the motion of a particle under a central acceleration; that is, one always directed towards or away from a fixed point, the exact law of the force of attraction as a function of the distance being left indeterminate. The planetary motions are of this description. In this case the differential equation of the motion is

$$\frac{d^2\mathbf{r}}{dt^2} = \mathbf{r}_1 f(\mathbf{r}). \tag{94}$$

As the acceleration is always along and therefore parallel to the radius vector, the product

$$\frac{d^2\mathbf{r}}{dt^2} \times \mathbf{r} = 0.$$

This may be written $\quad \dfrac{d}{dt}\left(\dfrac{d\mathbf{r}}{dt} \times \mathbf{r}\right) = 0,$

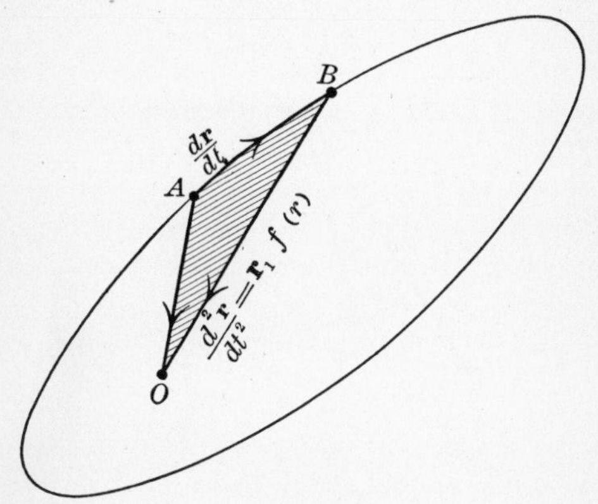

FIG. 47.

for, carrying out the differentiations, we obtain from this last

$$\frac{d^2\mathbf{r}}{dt^2} \times \mathbf{r} + \frac{d\mathbf{r}}{dt} \times \frac{d\mathbf{r}}{dt} = 0,$$

of which the second term on the left is zero, because any vector product containing parallel vectors is zero. Hence, integrating,

$$\frac{d\mathbf{r}}{dt} \times \mathbf{r} = \text{const. vector} = \mathbf{c}, \tag{95}$$

where $\mathbf{c}$ is a vector perpendicular to the plane of $\mathbf{r}$ and $\dfrac{d\mathbf{r}}{dt}$.

But $\frac{d\mathbf{r}}{dt}$ is the vector velocity along the tangent and $\mathbf{r}$ is the radius vector, so that by § 18 the above equation is *twice* the area swept out by the radius vector in unit time. We obtain the result, then, that under *any* central acceleration the rate of description of areas is a constant, and the orbit lies in a plane perpendicular to a constant vector $\mathbf{c}$.

Harmonic Motion. Equation of Ellipse. As another example to integrate

$$\frac{d^2\mathbf{r}}{dt^2} + m^2\mathbf{r} = 0, \tag{96}$$

which is the equation of a central acceleration proportional to the distance from the center. Such motions take place wherever **Hooke's Law** is followed.

We know that the two solutions of the scalar equation

$$\frac{d^2r}{dt^2} + m^2r = 0,$$

are $$r = a\cos mt \text{ and } r = b\sin mt,$$

and that the complete solution is the sum of these two. If we replace the arbitrary constants a and b by the arbitrary constant vectors $\mathbf{a}$ and $\mathbf{b}$, obtaining

$$\mathbf{r} = \mathbf{a}\cos mt + \mathbf{b}\sin mt, \tag{97}$$

it is easily seen, by differentiation, that this equation is the complete solution of the vector differential equation

$$\frac{d^2\mathbf{r}}{dt^2} + m^2\mathbf{r} = 0.$$

By an extension of this process, which is easily seen to hold, we may then state the rule for the solution of linear differential equations to any order with constant coefficients: *Find the solutions, assuming the vector variable to be a scalar variable, multiply these each by an arbitrary vector and add. The result will be the complete vector solution.*

These arbitrary vectors are to be determined from the initial or final conditions of the problem exactly in the same manner as we do with scalar equations. Equation (97) is the composition of two simple harmonic motions along directions determined by **a** and **b**, and is easily seen to represent in

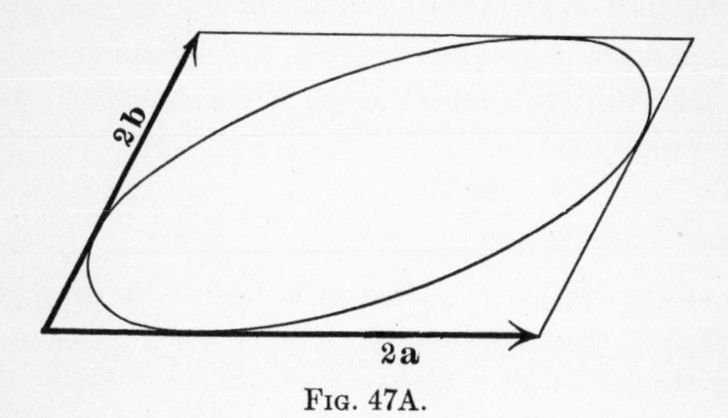

FIG. 47A.

general an ellipse inscribable in the parallelogram whose sides are 2 **a** and 2 **b** respectively.

$$\mathbf{r} = \mathbf{a} \cos mt + \mathbf{b} \sin mt$$

is therefore one form of the equation of an ellipse, if m is real.

42. Hodograph and Orbit under Newtonian Forces. As another example in vector differentiation and integration consider the case of motion under a force directed along the radius vector and inversely proportional to the *square* of the distance in magnitude. This is the ordinary planetary motion. We are to solve the differential equation

$$\frac{d^2\mathbf{r}}{dt^2} = \frac{m\,\mathbf{r}_1}{r^2}\,.\text{*} \tag{a}$$

Multiplying by $\mathbf{r}\times$, we have at once

$$\mathbf{r}\times \frac{d^2\mathbf{r}}{dt^2} = 0,$$

and hence

$$\mathbf{r}\times \frac{d\mathbf{r}}{dt} = \mathbf{c}, \tag{b}$$

* This equation states that the acceleration is directed outwards, i.e. the forces are repulsive. For attractive forces change m to –m.

where $\mathbf{c}$ is a constant vector. This last equation states that the rate of description of areas is constant, as above.

Lemma in Differentiation. Consider now the identity

$$\mathbf{r} \equiv r\,\mathbf{r}_1,$$

then
$$d\mathbf{r} = r d\mathbf{r}_1 + \mathbf{r}_1 dr,$$

hence multiplying by $\mathbf{r}_1\times$,

$$\mathbf{r}_1\times d\mathbf{r} = r\,\mathbf{r}_1\times d\mathbf{r}_1,$$

so that
$$\mathbf{r}_1\times d\mathbf{r}_1 = \frac{\mathbf{r}\times d\mathbf{r}}{r^2},$$

a result we might have written down immediately by similar triangles.

Multiply this last result by $\mathbf{r}_1\times$, obtaining from the left-hand side of equation

$$\mathbf{r}_1\times(\mathbf{r}_1\times d\mathbf{r}_1) = \mathbf{r}_1\,(\mathbf{r}_1\cdot d\mathbf{r}_1) - d\mathbf{r}_1\,(\mathbf{r}_1\cdot\mathbf{r}_1) = - d\mathbf{r}_1,$$

since $\mathbf{r}_1$ and $d\mathbf{r}_1$ are perpendicular and $\mathbf{r}_1\cdot\mathbf{r}_1 = 1$, so that

$$d\mathbf{r}_1 = -\frac{\mathbf{r}_1\times(\mathbf{r}\times d\mathbf{r})}{r^2}$$

and
$$\frac{d\mathbf{r}_1}{dt} = -\frac{\mathbf{r}_1\times\left(\mathbf{r}\times\dfrac{d\mathbf{r}}{dt}\right)}{r^2}.$$

Now the parenthesis on the right is a constant vector for central forces by (b), so that, by (a),

$$\frac{d^2\mathbf{r}}{dt^2}\times\mathbf{c} = m\frac{\mathbf{r}_1\times\mathbf{c}}{r^2} = - m\frac{d\mathbf{r}_1}{dt},$$

and integrating,

$$\frac{d\mathbf{r}}{dt}\times\mathbf{c} = \mathbf{d} - m\,\mathbf{r}_1, \qquad (c)$$

where $\mathbf{d}$ is a constant vector perpendicular to $\mathbf{c}$, as is seen by multiplying by $\mathbf{c}\cdot$

Multiplying equation (c) by $\frac{1}{\mathbf{c}}\times$ we obtain

$$\frac{1}{\mathbf{c}}\times\left(\frac{d\mathbf{r}}{dt}\times\mathbf{c}\right)=\frac{d\mathbf{r}}{dt}\,\mathbf{c}\cdot\frac{1}{\mathbf{c}}-\mathbf{c}\frac{d\mathbf{r}}{dt}\cdot\frac{1}{\mathbf{c}}=\frac{1}{\mathbf{c}}\times\mathbf{d}-m\frac{1}{\mathbf{c}}\times\mathbf{r}_1.$$

This becomes, since $\mathbf{c}\cdot\frac{1}{\mathbf{c}}=1$, and because by ($b$) $\frac{d\mathbf{r}}{dt}$ is normal to $\mathbf{c}\left(\text{or to }\frac{1}{\mathbf{c}}\right)$,

$$\frac{d\mathbf{r}}{dt}=\mathbf{c}^{-1}\times\mathbf{d}-m\mathbf{c}^{-1}\times\mathbf{r}_1=\frac{1}{\mathbf{c}}\times(\mathbf{d}-m\mathbf{r}_1), \qquad (d)$$

the equation to the hodograph.

Since $\mathbf{c}$ and $\mathbf{d}$ are constant vectors, $\mathbf{c}^{-1}\times\mathbf{d}$ is a constant vector normal to $\mathbf{c}$, and hence lies in the plane of the orbit; $m\,\mathbf{c}^{-1}\times\mathbf{r}_1$ is a vector constant in length, so that the extremity of $\frac{d\mathbf{r}}{dt}$ lies in a circle drawn around the point $\mathbf{c}^{-1}\times\mathbf{d}$ as center and in the plane of the orbit. This length, since $\mathbf{c}$ and $\mathbf{r}_1$ are perpendicular, is $\frac{m}{c}$.

To obtain the equation to the path multiply the equation of the hodograph by $\mathbf{r}\times$,

$$\mathbf{r}\times\frac{d\mathbf{r}}{dt}=\mathbf{c}=\mathbf{r}\times(\mathbf{c}^{-1}\times\mathbf{d})-m\,\mathbf{r}\times(\mathbf{c}^{-1}\times\mathbf{r}_1).$$

Expanding the two triple vector products and remembering (b),

$$\mathbf{c}=\mathbf{c}^{-1}\,\mathbf{r}\cdot\mathbf{d}-m\,r\,\mathbf{c}^{-1}.$$

Multiplying by $\mathbf{c}\cdot$,

$$\mathbf{c}^2=\mathbf{r}\cdot\mathbf{d}-m\,r.$$

So that

$$r=\frac{c^2}{d\cos\alpha-m}=\frac{-\frac{c^2}{m}}{1-\frac{d}{m}\cos\alpha},$$

the polar equation of a conic; the angle α being measured from the line d.

Comparing with

$$r = \frac{l}{1 - e \cos \alpha},$$

the general polar equation of a conic where the focus is the origin, where l is the semi latus rectum, and where e is the eccentricity, we find that the path of the orbit is a conic of eccentricity $\frac{d}{m}$; that **d** is along the major axis and that the magnitude of the major axis is $\frac{mc^2}{m^2 - d^2}$.*

43. Partial Differentiation. When any vector is a function of more than one scalar variable it can be differentiated partially with respect to each one, the remaining variables being considered constant during the differentiations. Such partial differential coefficients are written just as in ordinary scalar calculus as $\frac{\partial \mathbf{a}}{\partial x}, \frac{\partial \mathbf{q}}{\partial z}$, etc., where $x, z, \ldots$ are the independent scalar variables. The total change in **a** due to simultaneous changes in the variables $dx, dy, dz, \ldots$ is written

$$d\mathbf{a} = \frac{\partial \mathbf{a}}{\partial x} dx + \frac{\partial \mathbf{a}}{\partial y} dy + \frac{\partial \mathbf{a}}{\partial z} dz. \tag{98}$$

Symbolically we may write this as

$$d\mathbf{a} = \left(\frac{\partial}{\partial x} dx + \frac{\partial}{\partial y} dy + \frac{\partial}{dz} dz\right)\mathbf{a}, \tag{99}$$

where the expression in parentheses is to be considered as a differential operator to be applied to **a**, as in (78).

Origin of the Operator Del (∇). The operator (99) has the form of a scalar product, the two constituent vectors of which are

$$\left(\mathbf{i}\frac{\partial}{\partial x} + \mathbf{j}\frac{\partial}{\partial y} + \mathbf{k}\frac{\partial}{\partial z}\right) \text{ and } (\mathbf{i}\,dx + \mathbf{j}\,dy + \mathbf{k}\,dz) = d\mathbf{r}.$$

* See Appendix, p. 245, for Path Described by an Electron in a Uniform Magnetic Field.

If the single symbol ∇ (read del) be used to denote the first expression

$$\nabla \equiv \left(\mathbf{i}\,\frac{\partial}{\partial x} + \mathbf{j}\,\frac{\partial}{\partial y} + \mathbf{k}\,\frac{\partial}{\partial z}\right), \tag{100}$$

the equation may be written

$$d\mathbf{a} = (\nabla\cdot d\mathbf{r})\,\mathbf{a}. \tag{101}$$

We are thus led naturally to the consideration of the properties of this symbolic vector ∇.

PROBLEMS AND EXERCISES.

1. If $\mathbf{r}\cdot d\mathbf{r} = 0$, show that $r = \text{const.}$

If $\mathbf{r}\times d\mathbf{r} = 0$, show that $\mathbf{r}_1 = \textbf{const.}$

If $\mathbf{r}\cdot d\mathbf{r}\times d^2\mathbf{r} = 0$, show that $\mathbf{r}\times d\mathbf{r}$ has a fixed direction and that $\mathbf{r}$ is always parallel to a fixed plane.

2. Show that

$$dr = \mathbf{r}_1\cdot d\mathbf{r} = \frac{1}{\mathbf{r}_1}\cdot d\mathbf{r}.$$

and that

$$\frac{dr}{r} = \mathbf{r}^{-1}\cdot d\mathbf{r}.$$

3. Given a particle moving in a plane curve, in the plane of $\mathbf{ij}$, obtain the components of $\frac{d\mathbf{r}}{dt}$ along and perpendicular to the radius vector.

They are

$$\frac{dr}{dt}\,\mathbf{r}_1 \text{ and } r\,\frac{d\theta}{dt} \perp \mathbf{r}_1.$$

A unit line $\perp$ to $\mathbf{r}_1$ is $\mathbf{k}\times\mathbf{r}_1$, where $\mathbf{k}$ is normal to the plane.

4. Obtain similarly by differentiation of

$$\frac{d\mathbf{r}}{dt} = \frac{dr}{dt}\,\mathbf{r}_1 + \frac{d\theta}{dt}\,(\mathbf{k}\times\mathbf{r})$$

the accelerations along $\mathbf{r}_1$ and perpendicular to $\mathbf{r}_1$.

They are

$$\left\{\frac{d^2r}{dt^2} - r\left(\frac{d\theta}{dt}\right)^2\right\}\mathbf{r}_1 \text{ and } \left(r\frac{d^2\theta}{dt^2} + 2\frac{dr}{dt}\frac{d\theta}{dt}\right)\perp\mathbf{r}_1.$$

5. If r, ϕ, θ be a system of polar coördinates in space, where r is the distance of a point from the origin, ϕ the meridional angle, and θ the polar angle, obtain the expressions for the components of the velocity along the radius vector, the meridian, and a parallel of latitude.

6. Find the accelerations along the same directions in the problem above.

Express them in Cartesian form.

7. The curve

$$\mathbf{r} = \mathbf{a} \cos t + \mathbf{b} \sin t$$

represents an ellipse of which $\mathbf{a}$ and $\mathbf{b}$ are conjugate radii. The vector

$$\mathbf{r}' = \frac{d\mathbf{r}}{dt} = -\mathbf{a} \sin t + \mathbf{b} \cos t$$

$$= \mathbf{a} \cos\left(\frac{\pi}{2} + t\right) + \mathbf{b} \sin\left(\frac{\pi}{2} + t\right)$$

is the radius conjugate to $\mathbf{r}$, and parallel to the tangent at $\mathbf{r}$.

8. The parallelogram determined by the conjugate radii of an ellipse is constant in area.

$$[\mathbf{r} \times \mathbf{r}' = \text{const.}].$$

9. An elliptical helix is represented by

$$\mathbf{r} = \mathbf{a} \cos t + \mathbf{b} \sin t + \mathbf{c}t.$$

10. Show that the tangent line and the osculating plane of any curve $\mathbf{r} = \mathbf{f}(s)$ may be respectively written in the forms,

$$\boldsymbol{\rho} = \mathbf{r} + x\mathbf{r}',$$

$$\boldsymbol{\rho} = \mathbf{r} + x\mathbf{r}' + y\mathbf{r}'',$$

where $\mathbf{r}' \equiv \dfrac{d\mathbf{f}}{ds}$ and $\mathbf{r}'' \equiv \dfrac{d^2\mathbf{f}}{ds^2}$, and

x and y are variable scalars.

11. Find the tortuosity, $\mathbf{T}$, of any curve where $\mathbf{T}$ is defined as the rate change of the normal $\mathbf{n}$ to the osculating plane with respect to the arc ds.

$$\mathbf{T}_0 = \left(\frac{d\mathbf{n}}{ds}\right)_0 = \frac{\left(\dfrac{d\mathbf{r}}{ds} \times \dfrac{d^2\mathbf{r}}{ds^2} \cdot \dfrac{d^3\mathbf{r}}{ds^3}\right)}{\dfrac{d^2\mathbf{r}}{ds^2} \cdot \dfrac{d^2\mathbf{r}}{ds^2}}.$$

Express this in Cartesian notation.

12. Find the curvature of a circular helix. Find the tortuosity of a circular helix.

13. The equation

$$\mathbf{r} = x\,\boldsymbol{\phi}(t) + \mathbf{a},$$

where $\mathbf{a}$ is a constant vector, represents a cone standing on the curve $\mathbf{r} = \boldsymbol{\phi}(t)$ with its vertex at the extremity of $\mathbf{a}$.

14. The equation

$$\mathbf{r} = \boldsymbol{\phi}(t) + x\,\mathbf{a},$$

where $\mathbf{a}$ is a constant vector, represents a cylinder standing on the curve $\mathbf{r} = \boldsymbol{\phi}(t)$ and having its generators parallel to $\mathbf{a}$.

15. Prove that the acceleration of a particle moving in a circle with uniform speed is given by

$$\frac{d^2\mathbf{r}}{dt^2} = -\frac{v^2}{\mathbf{r}}.$$

16. Write out the equations of the hodograph for uniformly accelerated motion in a straight line.

17. Find the hodograph to the motion

$$\frac{d^2\mathbf{r}}{dt^2} = \pm\, m^2\mathbf{r}, \qquad \text{(a)}$$

where the acceleration varies as the distance from the origin.

The solution of (a) being that of $\left(\frac{d^2}{dt^2} \pm m^2\right)\mathbf{r} = 0$, that is

$$\mathbf{r} = \mathbf{A}\cos mt + \mathbf{B}\sin mt,$$

and

$$\mathbf{r} = \mathbf{A}e^{mt} + \mathbf{B}e^{-mt}.$$

Interpret these equations and those of the resulting hodographs.

18. Show that if the hodograph be a circle, and the acceleration be directed to a fixed point, the orbit must be a conic section, which is limited to being a circle if the acceleration follow any other law than the inverse square.

19. In the hodograph corresponding to acceleration $f(\mathbf{r})$ directed towards a fixed center, the curvature is inversely as $r^2 f(\mathbf{r})$.

20. Show directly without analysis that

$$\frac{1}{\mathbf{r}}\times d\mathbf{r} = \frac{1}{\mathbf{r}_1}\times d\mathbf{r}_1 = \mathbf{r}_1\times d\mathbf{r}_1,$$

and hence that

$$\frac{d\mathbf{r}_1}{dt} = \frac{-\,\mathbf{r}_1\times\left(\mathbf{r}\times\frac{d\mathbf{r}}{dt}\right)}{r^2}.$$

CHAPTER V.

THE DIFFERENTIAL OPERATORS.

The Vector Operator ∇ (read del). This sign is sometimes called "nabla" (Heaviside) and also "atled," which is "delta" (Δ) reversed. The term "**del**" is, however, well worthy of adoption, as it is short, easy to pronounce and conflicts with no other terminology. As ∇ is the most important differential operator in mathematical physics its properties will be studied in detail.

Definition. ∇ is defined by the equation

$$\nabla \equiv \mathbf{i}\frac{\partial}{\partial x} + \mathbf{j}\frac{\partial}{\partial y} + \mathbf{k}\frac{\partial}{\partial z}. \tag{102}$$

We have already come across the scalar differential operator p on page 72. The paragraphs concerning p should be consulted at this point. As by its definition ∇ is made up of three symbolic components along the three axes **i j k**, the symbolic magnitudes of them being $\frac{\partial}{\partial x}$, $\frac{\partial}{\partial y}$, and $\frac{\partial}{\partial z}$ respectively, it may be looked upon as a symbolic vector itself. This view of ∇ as a vector, is important and of great help in the comprehension of what follows. The employment of ∇ in the treatment of the physical properties of space is of the most frequent occurrence. It is, therefore, extremely desirable to have a geometric or visual representation of such physical properties in space, or fields, as they are called, and of the effect of operators upon them.

44. Scalar and Vector Fields. Reconsult § 5 at this point.

Definition. If to every point in a region, finite or not, there corresponds a definite value of some physical property, the region so defined is called a *field*. Should this property be a scalar one the field is called a scalar field. As examples of such may be mentioned the temperature at any given instant, at all points of a body; or the density at all points; or the potential at all points, due to electrical, magnetic, or gravitational matter respectively.

On the other hand, if the property is a vector one it is said to be a vector field. As examples of these are the velocity at all points of a fluid; the electrical, magnetic, or gravitational intensity (of force) at all points of a region due to electrical, magnetic, or gravitational matter, respectively.

45. Scalar and Vector Functions of Position. Assume any arbitrary origin O and from it draw a variable radius vector $\mathbf{r}$. This vector $\mathbf{r}$ may extend to and determine any point in space. By the term "value of $\mathbf{r}$" is meant the "position of the terminus of $\mathbf{r}$." Now, if to every value of $\mathbf{r}$ there corresponds a definite scalar quantity V, V is said to be a scalar point-function of $\mathbf{r}$ and is written

$$V = f(\mathbf{r}). \tag{103}$$

If to every value of $\mathbf{r}$ there corresponds a definite vector quantity $\mathbf{F}$, $\mathbf{F}$ is said to be a vector point-function of $\mathbf{r}$ and is written

$$\mathbf{F} = \mathbf{f}(\mathbf{r}). \tag{104}$$

$V = f(\mathbf{r})$ and $\mathbf{F} = \mathbf{f}(\mathbf{r})$ are thus the functional representations of scalar and vector fields respectively.

Mathematical and Physical Discontinuities. The functions met with in physics are almost always continuous and single-valued except perhaps at isolated points, lines or surfaces finite in number. If not, they can be made so by various devices, such as by inserting diaphragms to prevent passing into a region by two or more different paths, etc.

The functions dealt with, in what follows, are supposed to be of this description.

The most common kinds of discontinuities that occur in mathematics are those in which, either the value itself of a function suffers an abrupt change, or where the rate of change of the function abruptly takes on a new value as the independent variable is continuously increased or diminished.

Graphically these mean a break in the curve, or a sudden change in the direction of the curve representing the function, as in Fig. 48 (a), at P and at Q. In nature such discontinuities do *not* take place. For example, the temperature cannot have one value on one side of a surface and another value on the other side, where the two sides of the surface are infinitely near to each other. In reality there is a *continuous* but very rapid change in the temperature from its value on one side to its value on the other as we pass through the surface. Besides, infinitely thin surfaces do not exist except in our imagination.

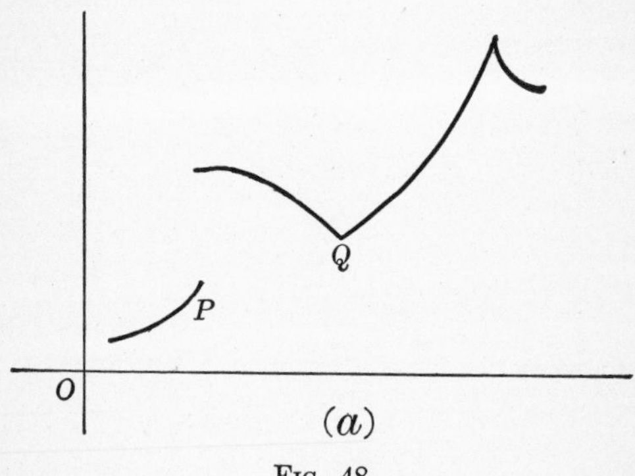

FIG. 48.

If the temperature gradient in a body has one definite value and seems to change abruptly to another value quite different from the first, we know that in reality there is a very rapid but finite rate of change of the gradient at the place in question. This absence of discontinuity in any natural function is indicated in Fig. 48 (b), which shows the continuous function, which to all intents and purposes replaces the discontinuous function (a). It is for this reason that the usual attention will not be paid to the consideration of discontinuities in what follows. All natural functions being in reality continuous, such consideration is physically superfluous. We may state the same idea explained above, by saying that on sufficient magnification of finite amount, all curves representing natural phenomena will be found to be continuous.

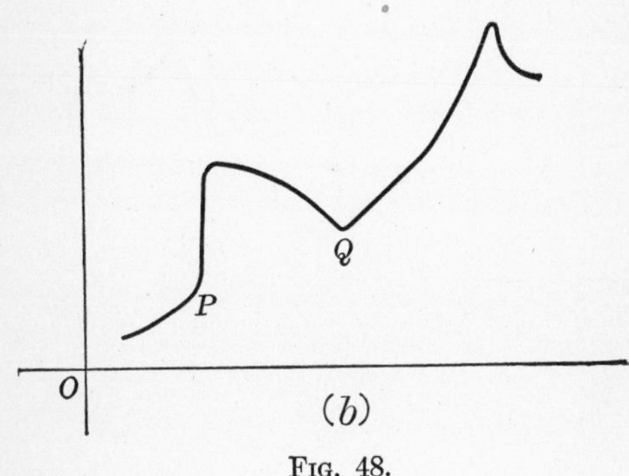

Fig. 48.

We do not, however, wish to convey the idea that the *study* of discontinuities is unimportant, as on the contrary the mathematical results derived from their study are of the greatest importance, and teach us how to attack problems involving sudden natural changes, or as we might call them, "apparent discontinuities" in physical functions. In fact the methods generally employed are to *assume* them to be actual mathematical discontinuities and treat them as such. The point we wish to make is that in the *general* analytical expression of natural phenomena it is unnecessary to complicate the formulæ by the separate consideration of discontinuities, but to let the student treat them by the recognized mathematical methods whenever it is convenient or necessary to do so.

46. The Potential. For the sake of definiteness we shall consider the potential due to *electrical* matter. The whole argument applies almost identically to magnetic or gravitational potential; to the distribution of temperature in a body, or to the velocity-potential in moving fluids, etc., etc.

Definition. The potential at any point in space due to a distribution of electrical matter may be defined as the work done on a unit positive quantity of electricity as it is brought *by any path* from infinity to that point. As like charges repel each other, it will require positive work to be done *on* the unit charge to bring it in the neighborhood of any positive distribution of electricity, and hence the potential around such a distribution will be positive, *increasing* as the points are taken nearer and nearer to it. It is evident, also, that the forces acting on the unit charge are repelling forces and that they act in the direction *opposite* to the increase of potential.

For instance in the electric field due to the charge $+ q$ on a small sphere, the unit positive charge at P is repelled by a force acting radially outward, of amount calculated by Coulomb's Law

$$\mathbf{F} = \frac{q}{r^2} \mathbf{r}_1,$$

where r is the distance from P to the center of the sphere. The force $\mathbf{F}$ evidently becomes greater as P approaches the sphere, and work has to be done upon the unit charge in order to make it do so.

Level or Equipotential Surfaces. Let all the points having the same potential be found, or, in other words, find all those points which require the same amount of work to be expended upon the unit positive charge to bring it from infinity up to them. If C be this amount of work and V be the potential function, then the equation to the locus defined by these points will be

$$V(\mathbf{r}) = C,$$

where $\mathbf{r}$ is measured from some arbitrary origin.

Find similarly all points which require a small amount more of work $C + dC$; the equation to this locus will be

$$V(\mathbf{r}) = C + dC.$$

In the special case of the sphere, Fig. 48A, these points will lie on spheres concentric with the charged sphere.

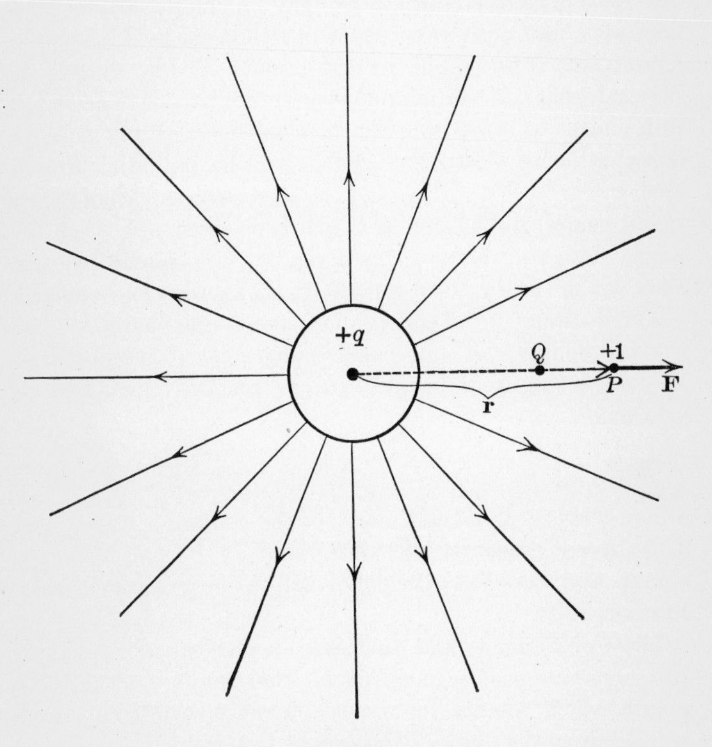

FIG. 48A.

These equations define surfaces which are called Level or Equipotential Surfaces of the function $V(\mathbf{r})$. Let many such surfaces be constructed and let the quantities of work employed in reaching the successive ones differ by *equal* amounts.

It requires no work to carry the unit charge from one point to another having the same potential, for by definition it requires the same amount of work to bring the unit charge from infinity to either of these points by any path, and we may choose the path leading to the second point to pass

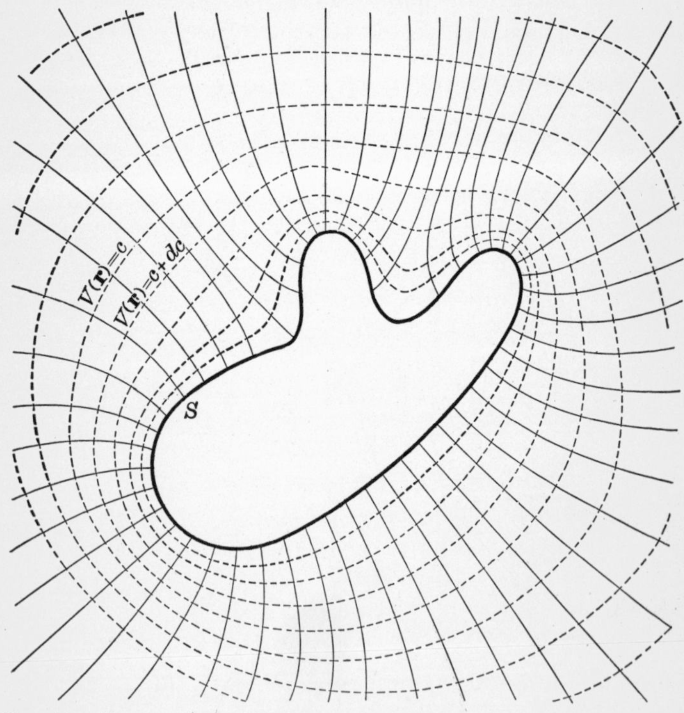

FIG. 49. Showing Lines of Force and Equipotential Surfaces Around a Charged Conductor.

through the first point. Hence the work done in going from the first point to the second point must be zero.

Relation between Force and Potential. Consider in particular two adjacent level surfaces, the difference in potential between them being dV. This means that it requires

dV units of work to carry the unit charge from one of the surfaces to the other in *any* manner.

Since the amount of work is constant in going from one level surface to the next one, the greatest forces will be encountered in going by the shortest path from one to the other, so that at any *given point* P_1 the maximum force is along the common perpendicular to the two surfaces. Com-

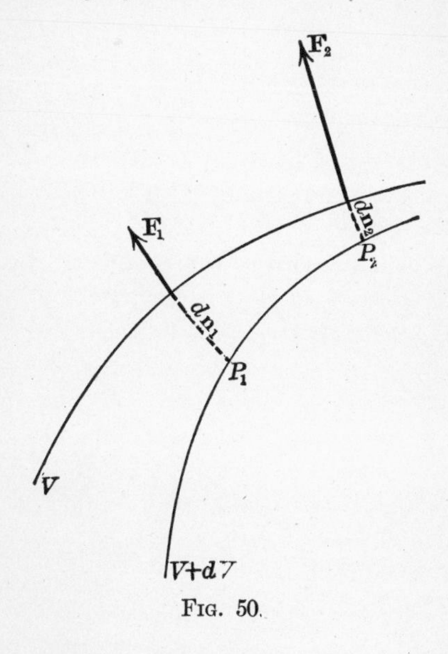

FIG. 50.

paring the forces at *different points* P_1 or P_2 these maximum forces $\mathbf{F}_1$ or $\mathbf{F}_2$ will be found greater the nearer the surfaces are together. Hence the forces in the field are normal to the level surfaces and are inversely proportional to their distance apart. But V increases most rapidly along the normal to a level surface and its rate of increase is greatest where the surfaces lie closest together. We are therefore led to expect a relation between the force at a point and the rate of increase of V at the same point. All this is concisely represented by writing for the work done in going from one surface to the next,

$$dV = \mathbf{F} \cdot d\mathbf{n},$$

where $d\mathbf{n}$ is the normal distance between the two surfaces, so that

$$\mathbf{F} = \frac{dV}{dn} (-\mathbf{n}). \tag{105}$$

where $\mathbf{n}$ is the unit normal pointing in the direction of *increasing* potential. This important equation states that the force at any point is normal to the level surface passing through that point; opposite to the direction of fastest increase in V, and equal in magnitude to this fastest rate of increase.

Thus a knowledge of the potential everywhere gives a knowledge of the forces everywhere *not only in magnitude but in direction as well.*

A scalar point-function, as it does not involve direction, is clearly simpler of representation on a diagram than a vector one. The potential function is very useful for this reason, as a complete knowledge of its value everywhere immediately gives us a complete knowledge of the forces everywhere. Thus the comparatively simple scalar function intrinsically contains all that we wish to know about the comparatively more complicated vector function. This property alone is sufficient to justify its invention and use.

47. ∇ Applied to a Scalar Point-Function. Gradient or Slope of a Scalar Point-Function.

Definition. The *vector*, perpendicular to the level surface at any point, equal in magnitude to the fastest rate of increase of V, and pointing in the direction of this fastest *increase*, is called the **gradient** or the *slope* of V at that point and is written

$$\text{grad } V \quad \text{or} \quad \text{slope } V,$$

preferably the first.

Grad V Independent of Choice of Axes. The force **F** acting on the unit charge is, by the above definition in connection with § 46, evidently equal to $-$ grad V, but as **F** is entirely independent of any choice of axes, so is $-$ grad V independent of them.

It remains to show that the operator ∇ applied to V gives

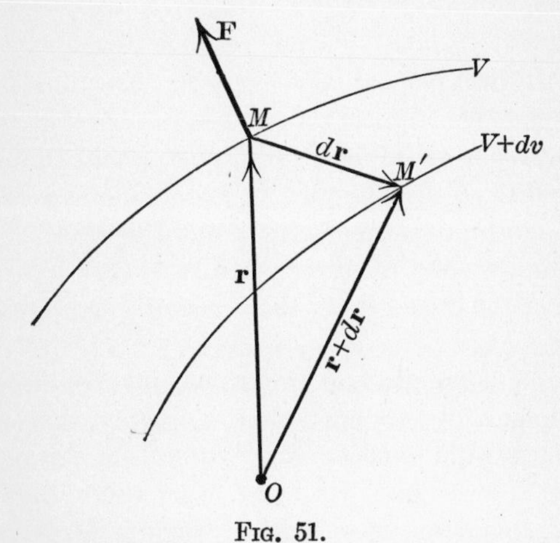

FIG. 51.

the grad as defined above. The work done on a unit charge as it is carried from M to M' is, by § 16, where $\overrightarrow{MM'} = d\mathbf{r}$,

$$-\mathbf{F}\cdot d\mathbf{r} = \text{grad } V\cdot d\mathbf{r} = dV.$$

Considering now V as a function of $x\ y\ z$,

$$dV = \frac{\partial V}{\partial x}\,dx + \frac{\partial V}{\partial y}\,dy + \frac{\partial V}{\partial z}\,dz,$$

so that

$$\text{grad } V\cdot d\mathbf{r} = \left(\mathbf{i}\,\frac{\partial V}{\partial x} + \mathbf{j}\,\frac{\partial V}{\partial y} + \mathbf{k}\,\frac{\partial V}{\partial z}\right)\cdot(\mathbf{i}\,dx + \mathbf{j}\,dy + \mathbf{k}\,dz)$$

$$= \nabla V\cdot d\mathbf{r}.$$

As this equation is true whatever path $d\mathbf{r}$ is taken between the two surfaces,

$$\text{grad } V \equiv \nabla V. \tag{106}$$

Thus the application of the operator ∇ to a scalar point function is a vector which gives its rate of most rapid increase in magnitude and direction. The significance and importance of this operator is now easily understood.

The vector ∇V is often called after **Lamé*** the first differential parameter of V.

Fourier's Law. If instead of potential we consider temperature, the level surfaces are then isothermal surfaces, and the ∇ of the temperature function gives the rate of the most rapid increase of the temperature in magnitude and direction. As the flow of heat takes place in the direction of most rapid *decrease*, $\mathbf{q}$, the intensity of flow, is given by

$$\mathbf{q} = -k\nabla\theta,$$

where k is a characteristic of the medium at the point in question and called its conductivity, and θ is the temperature at any point. This is called Fourier's Law for the flow of heat.

48. Illustrations of the Application of ∇ to Scalar Functions of Position. By means of ordinary partial differentiation on the functions

$$r = (x^2 + y^2 + z^2)^{\frac{1}{2}}$$

and

$$r^n = (x^2 + y^2 + z^2)^{\frac{n}{2}}$$

the following important results are obtained:

$$\nabla r = \left(\mathbf{i}\frac{\partial}{\partial x} + \mathbf{j}\frac{\partial}{\partial y} + \mathbf{k}\frac{\partial}{\partial z}\right)(x^2 + y^2 + z^2)^{\frac{1}{2}}$$

$$= \frac{x\,\mathbf{i} + y\,\mathbf{j} + z\,\mathbf{k}}{(x^2 + y^2 + z^2)^{\frac{1}{2}}} = \frac{\mathbf{r}}{r} = \mathbf{r}_1$$

* G. Lamé. Leçons sur les coordonneés curvilignes et leurs diverses applications. Paris, 1859.

and

$$\nabla r^n = \left(\mathbf{i}\,\frac{\partial}{\partial x} + \mathbf{j}\,\frac{\partial}{\partial y} + \mathbf{k}\,\frac{\partial}{\partial z}\right)(x^2 + y^2 + z^2)^{\frac{n}{2}}$$

$$= n\,(x^2 + y^2 + z^2)^{\frac{n}{2}-1}\,(x\,\mathbf{i} + y\,\mathbf{j} + z\,\mathbf{k})$$

$$= n\,(x^2 + y^2 + z^2)^{\frac{n-1}{2}}\,\frac{x\,\mathbf{i} + y\,\mathbf{j} + z\,\mathbf{k}}{(x^2 + y^2 + z^2)^{\frac{1}{2}}}$$

$$= nr^{n-1}\nabla r = nr^{n-1}\,\mathbf{r}_1 = nr^{n-2}\,\mathbf{r}. \qquad (108)$$

So that ∇ differentiates the function r^n similarly to the scalar differentiation of u^n by $\frac{d}{dx}$,

$$\frac{du^n}{dx} = nu^{n-1}\,\frac{du}{dx}.$$

A shorter method for obtaining this formula will be given in § 49.

The particular cases

$$\nabla r = \mathbf{r}_1 \quad \text{and} \quad \nabla\,\frac{1}{r} = -\,\frac{\mathbf{r}_1}{r^2} \qquad (109)$$

are important results.

Application of ∇ to a Scalar Product. It is frequently necessary to apply ∇ to a scalar product such as **a·b.** Remembering the laws for the ordinary differentiation of scalar products and the definition of ∇ it is easily shown that

$$\nabla\,(\mathbf{a}\cdot\mathbf{b}) = \nabla_a(\mathbf{a}\cdot\mathbf{b}) + \nabla_b(\mathbf{a}\cdot\mathbf{b}),$$

where the subscript affixed to the ∇ shows which vector is considered *variable, i.e.*, the one it acts upon.

The notation

$$\nabla_a(\mathbf{a}\cdot\mathbf{b}) \equiv \nabla(\mathbf{a}\cdot\mathbf{b})_b \qquad (110)$$

is also sometimes used, where the subscript in the second term shows which vector is considered *constant* during the differentiation. $\nabla_a(\quad)$, then, from this point of view differentiates partially, the vector **a** alone being considered variable in the parenthesis. In particular, if the vector $\boldsymbol{\omega}$, say, of the scalar product $\boldsymbol{\omega}\cdot\mathbf{r}$ is a constant vector, we obtain the important result

$$\nabla_r(\boldsymbol{\omega}\cdot\mathbf{r}) = \boldsymbol{\omega}, \tag{111}$$

for

$$\begin{aligned}\nabla(\boldsymbol{\omega}\cdot\mathbf{r}) &= \nabla_r(\boldsymbol{\omega}\cdot\mathbf{r}) = \nabla_r(\omega_1 x + \omega_2 y + \omega_3 z) \\ &= \mathbf{i}\,\omega_1 + \mathbf{j}\,\omega_2 + \mathbf{k}\,\omega_3 = \boldsymbol{\omega}.\end{aligned}$$

49. The Scalar Operator $\mathbf{s}_1\cdot\nabla$ or Directional Derivative. Since ∇V is a vector, its scalar product with any other vector $\mathbf{s}_1$ may be taken, and by definition this would give the component of the magnitude of ∇V in the direction of that vector. This is ordinarily written

$$\frac{\partial V}{\partial s} = s_1\frac{\partial V}{\partial x} + s_2\frac{\partial V}{\partial y} + s_3\frac{\partial V}{\partial z},$$

where s_1, s_2, s_3 are the components of the unit vector $\mathbf{s}_1$, or what is the same thing, the direction cosines of the direction $\mathbf{s}_1$.

This expression may be looked upon as the operation of

$$\mathbf{s}_1\cdot\nabla \equiv s_1\frac{\partial}{\partial x} + s_2\frac{\partial}{\partial y} + s_3\frac{\partial}{\partial z} = \frac{d}{ds} \tag{112}$$

upon V, which is the familiar directional derivative of V in the direction $\mathbf{s}_1$ or, otherwise,

$$(\mathbf{s}_1\cdot\nabla)V = \mathbf{s}_1\cdot\nabla V \quad \left(= \frac{d}{ds}V\right), \tag{113}$$

so that the directional derivative in any direction is the component of the magnitude of the gradient in that direction. The directional derivative then applied to the potential function gives the component of the force in the direction in which it is taken.

Total Derivative of a Function. If $\mathbf{s}_1$ be replaced by the small vector $d\mathbf{r}$, the operator $d\mathbf{r}\cdot\nabla$ applied to a scalar point function gives the total derivative of the function because

$$d\mathbf{r}\cdot\nabla V = (d\mathbf{r}\cdot\nabla)\,V = \left(\frac{\partial}{\partial x}\,dx + \frac{\partial}{\partial y}\,dy + \frac{\partial}{\partial z}\,dz\right)V = dV. \tag{114}$$

This is the same thing as the directional derivative along $(d\mathbf{r})_1$ multiplied by dr, or

$$d\mathbf{r}\cdot\nabla V = dr\,((d\mathbf{r})_1\cdot\nabla V). \tag{115}$$

We may now obtain equation (108) for the differentiation of r^n by means of the identity (114)

$$d\mathbf{r}\cdot\nabla\,(\quad) \equiv d_r\,(\quad),$$

as
$$d_r r^n = n r^{n-1}\,dr = n r^{n-1}\,\mathbf{r}_1\cdot d\mathbf{r},$$

so that the factor of $\cdot d\mathbf{r}$, $n r^{n-1}\,\mathbf{r}_1$ is the ∇r^n

and
$$\nabla r^n = n r^{n-1}\,\mathbf{r}_1 = n r^{n-2}\mathbf{r}$$

as before.

50. The Scalar Operator $\mathbf{s}_1\cdot\nabla$ Applied to a Vector. The operator $\mathbf{s}_1\cdot\nabla$ may be also applied to a vector point-function $\mathbf{F}$, giving as a result a new vector function, thus:

$$\begin{aligned}(\mathbf{s}_1\cdot\nabla)\,\mathbf{F} &= \mathbf{s}_1\cdot\nabla\,(\mathbf{i}F_1 + \mathbf{j}F_2 + \mathbf{k}F_3)\\ &= \mathbf{i}\ \mathbf{s}_1\cdot\nabla F_1 + \mathbf{j}\ \mathbf{s}_1\cdot\nabla F_2 + \mathbf{k}\ \mathbf{s}_1\cdot\nabla F_3\\ &= \mathbf{i}\left(s_1\,\frac{\partial F_1}{\partial x} + s_2\frac{\partial F_1}{\partial y} + s_3\,\frac{\partial F_1}{\partial z}\right)\\ &\quad + \mathbf{j}\left(s_1\,\frac{\partial F_2}{\partial x} + s_2\frac{\partial F_2}{\partial y} + s_3\,\frac{\partial F_2}{\partial z}\right)\\ &\quad + \mathbf{k}\left(s_1\,\frac{\partial F_3}{\partial x} + s_2\frac{\partial F_3}{\partial y} + s_3\frac{\partial F_3}{\partial z}\right).\end{aligned} \tag{116}$$

This is the directional derivative of the vector function $\mathbf{F}$ in the direction $\mathbf{s}_1$. It is also the vector whose components are the directional derivatives of the components of $\mathbf{F}$.

The parentheses may be omitted in $\mathbf{s}_1\cdot\nabla(\mathbf{F})$ and the expression written simply $\mathbf{s}_1\cdot\nabla\mathbf{F}$, but it does not mean that $(\mathbf{s}_1\cdot\nabla)\mathbf{F} \equiv \mathbf{s}_1\cdot(\nabla\mathbf{F})$, because $\nabla\mathbf{F}$ has no significance in this analysis. $\mathbf{s}_1\cdot\nabla\mathbf{F}$ may then be interpreted as nothing else but $(\mathbf{s}_1\cdot\nabla)\,\mathbf{F}$.

We may prove that the component along any constant vector $\mathbf{a}$ of the directional derivative along $\mathbf{b}$, of a vector function $\mathbf{F}$, is the directional derivative along $\mathbf{b}$ of the component of $\mathbf{F}$ along $\mathbf{a}$; that is,

$$\mathbf{a}_1\cdot(\mathbf{b}_1\cdot\nabla_F\mathbf{F}) = \mathbf{b}_1\cdot\nabla_F(\mathbf{a}_1\cdot\mathbf{F}).$$

And even without restriction to unit vectors that

$$\mathbf{a}\cdot(\mathbf{b}\cdot\nabla_F\mathbf{F}) = \mathbf{b}\cdot\nabla_F(\mathbf{a}\cdot\mathbf{F}). \tag{117}$$

This follows directly because ∇ differentiating $\mathbf{F}$ alone, $\mathbf{a}\cdot$ may be placed after the ∇. Also because (116)

$$\begin{aligned}
\mathbf{a}\cdot(\mathbf{b}\cdot\nabla\mathbf{F}) &= \mathbf{a}\cdot\mathbf{i}\;\mathbf{b}\cdot\nabla F_1 + \mathbf{a}\cdot\mathbf{j}\;\mathbf{b}\cdot\nabla F_2 + \mathbf{a}\cdot\mathbf{k}\;\mathbf{b}\cdot\nabla F_3\\
&= a_1\mathbf{b}\cdot\nabla F_1 + a_2\mathbf{b}\cdot\nabla F_2 + a_3\mathbf{b}\cdot\nabla F_3\\
&= \mathbf{b}\cdot\nabla_F(a_1F_1 + a_2F_2 + a_3F_3)\\
&= \mathbf{b}\cdot\nabla_F(\mathbf{a}\cdot\mathbf{F}).
\end{aligned}$$

Applying $\boldsymbol{\omega}\cdot\nabla$ to $\mathbf{r}$, the radius vector, gives

$$\begin{aligned}
(\boldsymbol{\omega}\cdot\nabla)\mathbf{r} &= \left(\omega_1\frac{\partial}{\partial x} + \omega_2\frac{\partial}{\partial y} + \omega_3\frac{\partial}{\partial z}\right)(\mathbf{i}\,x + \mathbf{j}\,y + \mathbf{k}\,z)\\
&= \mathbf{i}\,\omega_1\frac{\partial x}{\partial x} + \mathbf{j}\,\omega_2\frac{\partial y}{\partial y} + \mathbf{k}\,\omega_3\frac{\partial z}{\partial z}\\
&= \mathbf{i}\,\omega_1 + \mathbf{j}\,\omega_2 + \mathbf{k}\,\omega_3 = \boldsymbol{\omega}.
\end{aligned}$$

This expression should not be confounded with $\boldsymbol{\omega}\cdot\nabla r$, where the r is the *magnitude* of $\mathbf{r}$. Since $\nabla r = \mathbf{r}_1$ the value of this last expression would be

$$\boldsymbol{\omega}\cdot\nabla r = \boldsymbol{\omega}\cdot\mathbf{r}_1.$$

Combining $\boldsymbol{\omega}\cdot\nabla\mathbf{r} = \boldsymbol{\omega}$ with equation (111), we see that

$$(\boldsymbol{\omega}\cdot\nabla)\,\mathbf{r} = \nabla_r\,(\boldsymbol{\omega}\cdot\mathbf{r}) = \boldsymbol{\omega}. \tag{118}$$

The Operation ∇ on a Vector Point-Function. Any vector point-function **F** may be resolved into three components along **i j** and **k** so that

$$\mathbf{F} = \mathbf{i}\, F_1 + \mathbf{j}\, F_2 + \mathbf{k}\, F_3.$$

F_1, F_2, and F_3 are scalar functions of $x\ y\ z$, or of **r**. Considering ∇ as a vector, the product $\nabla\mathbf{F}$ can have no meaning unless the definition of the product of two vectors **a** and **b** be extended so that the product **ab** (without dot or cross) shall have a meaning.* But the scalar product of ∇ and a vector **F** and the vector product of ∇ and a vector **F** may be found by rules already given. The two expressions $\nabla\cdot\mathbf{F}$ and $\nabla\times\mathbf{F}$ are of such importance that special names have been given to them.

51. Divergence. The operator $\nabla\cdot(\ \ )$ or div ( ) [read del dot ( ) or divergence of ( )] when applied to the function **F** gives a *scalar* which in Cartesian notation is

$$\nabla\cdot\mathbf{F} \equiv \left(\mathbf{i}\frac{\partial}{\partial x} + \mathbf{j}\frac{\partial}{\partial y} + \mathbf{k}\frac{\partial}{\partial z}\right)\cdot(\mathbf{i}F_1 + \mathbf{j}F_2 + \mathbf{k}F_3)$$

$$\equiv \frac{\partial F_1}{\partial x} + \frac{\partial F_2}{\partial y} + \frac{\partial F_3}{\partial z} \equiv \text{div }\mathbf{F}. \qquad (119)$$

In order to obtain a physical interpretation of this quantity consider any vector field, the field of force due to an electrical distribution for instance. The convention usually adopted is that from every unit positive charge there originate 4π lines of force and into every negative unit charge there end 4π lines of force. The exact number of lines of force that issue from unit charge which convention has adopted, is of absolutely no consequence in this argument, and the new system which assumes the unit charge to give rise to but one

* This has not been done in this book, although Professor Gibbs has achieved beautiful results in his researches using this extended definition. The product **ab** he calls a *dyad*. See Gibbs-Wilson, Vector Analysis, Chapter V.

line may be adopted if desirable. If an element of volume be considered, for example a small parallelopiped with its sides $dx\ dy\ dz$ parallel to the axes of $x\ y$ and z respectively, the amount of electrical matter within it may evidently be measured by finding the excess of the lines which come out of it, over those which go into it. For every unit of positive

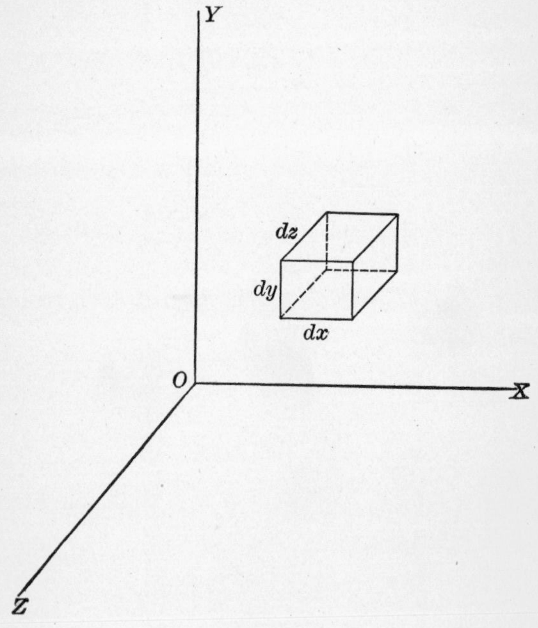

FIG. 52.

electricity there would emanate 4π lines outward, and for every negative unit 4π lines would enter into it. Considering these lines to cancel each other when going in opposite directions it is easily seen that the algebraic sum of the charges within the box may be found in amount and sign by an examination of the lines which leave or enter the box. Hence the lines which *diverge* from the element will be a measure of the positive charge within it. If the charge is negative, lines will end inside of the box, and therefore will *converge* into it.

To obtain an analytical expression for this quantity resolve **F**, or the flux of force, as it is called, into its three components parallel to **i j** and **k**. The flux into the face parallel to the yz-plane nearest the origin is $F_1\, dy\, dz$, the flux out of the opposite face is

$$\left(F_1 + \frac{\partial F_1}{\partial x} dx\right) dy\, dz,$$

so that the amount which comes out in excess of that which goes in, as far as the x-component of **F** is concerned, is

$$\left(F_1 + \frac{\partial F_1}{\partial x} dx\right) dy\, dz - F_1\, dy\, dz = \frac{\partial F_1}{\partial x}\, dx\, dy\, dz.$$

Similarly, for the other two components, which are obtained in the same manner,

$$\frac{\partial F_2}{\partial y}\, dx\, dy\, dz,$$

$$\frac{\partial F_3}{\partial z}\, dx\, dy\, dz,$$

so that the total amount of the flux **F** which diverges from the box $dx\ dy\ dz$ is

$$\left(\frac{\partial F_1}{\partial x} + \frac{\partial F_2}{\partial y} + \frac{\partial F_3}{\partial z}\right) dx\, dy\, dz.$$

Dividing by $dx\, dy\, dz$, the element of volume, to obtain the amount of flux which would come out of a unit volume under the same conditions, there remains precisely

$$\operatorname{div} \mathbf{F} = \nabla\cdot\mathbf{F} = \frac{\partial F_1}{\partial x} + \frac{\partial F_2}{\partial y} + \frac{\partial F_3}{\partial z}. \qquad (120)$$

Strictly then the term *divergence* means the number of lines which diverge per unit volume.

If the operator $\nabla\cdot$ be applied to the vector function representing the flux of heat or the velocity of a fluid, it will give by exactly the same reasoning the rate at which heat is issuing from a point per unit volume or the rate at which the fluid is originating at a point per unit volume. In the case of heat, if the divergence exists and is positive, there must be at the point in question a *source* of heat, heat actually created, or else at the point where the heat is leaving the temperature must be diminishing.

In the case of fluids, if the divergence exists and is positive, there must be either a *source* of fluid, fluid actually created, or else the density of the body at the point must be diminishing. If the divergence is negative, the opposite conditions hold in both the above examples. For instance, if the divergence of heat is negative, or in other words, if it converges, there must be a *sink* of heat, heat actually destroyed, annihilated, or else the temperature at the place must be rising, etc. In the case of electricity, the existence of a positive divergence proves the existence of positive electrical matter at the point. The negative of divergence is sometimes called *convergence*. It is better, however, to retain but one of the terms and use the negative sign to indicate convergence.

52. The Divergence Theorem.* This important theorem has a significance almost axiomatic when considered in the light of the foregoing. Consider any closed surface S, Fig. 24, lying in a vector field **q**, the velocity, say, in a moving incompressible fluid. It is evident that the excess of fluid which comes out, over that which goes in, may be measured in two distinct ways: first, by finding the total outward normal flux over the surface, or second, by going throughout the interior and taking the algebraic sum of the sources and sinks or diver-

* For a rigid mathematical proof see A. G. Webster, Electricity and Magnetism, pp. 60–62; Dynamics, pp. 340–342; also R. Gans, Einführung in die Vektoranalysis, pp. 29–33. See Appendix, p. 252, for other theorems analogous to the Divergence Theorem. Also P. Appell, Traité de Mécanique rationnelle, tome III, p. 2.

gences for every infinitesimal volume element contained within the surface. In symbols this is most conveniently expressed as

$$\int\int_S \mathbf{n}\cdot\mathbf{q}\, dS = \int\int\int \nabla\cdot\mathbf{q}\, dv, \qquad (121)$$

where **n** is the *outward*-drawn unit normal, dS the element of surface, dv the element of volume.

In words this reads: In a vector field the surface integral of the outward flux (*i.e.*, normal component of flux, see § 17) over any closed surface S is equal to the volume integral of the divergence taken throughout the volume enclosed by S. This is the divergence theorem.

From a mathematical point of view this demonstration may not be considered rigorous, but the ideas that this interpretation gives should be clearly understood by every student. In Cartesian dress this theorem becomes

$$\int\int_S [q_1 \cos(nx) + q_2 \cos(ny) + q_3 \cos(nz)]\, dS$$

$$= \int\int\int_S \left(\frac{\partial q_1}{\partial x} + \frac{\partial q_2}{\partial y} + \frac{\partial q_3}{\partial z}\right) dx\, dy\, dz. \qquad (122)$$

The idea of divergence is evidently independent of any choice of axes since none are required for its conception. Considered as the result of operating by $\nabla\cdot$ it is invariant to change of axes because ∇ has been shown to be invariant. Its invariant character may also be directly proved, as usual, by a transformation of axes, but this is a long and unnecessary process.

Examples. In particular

$$\nabla\cdot\mathbf{r} = \left(\mathbf{i}\,\frac{\partial}{\partial x} + \mathbf{j}\,\frac{\partial}{\partial y} + \mathbf{k}\,\frac{\partial}{\partial z}\right)\cdot(\mathbf{i}\, x + \mathbf{j}\, y + \mathbf{k}\, z)$$

$$= \frac{\partial x}{\partial x} + \frac{\partial y}{\partial y} + \frac{\partial z}{\partial z} = 1 + 1 + 1 = 3. \qquad (123)$$

Let us apply this result, using $\mathbf{r}$ for $\mathbf{q}$ in the divergence theorem. We obtain immediately

$$\iint \mathbf{n} \cdot \mathbf{r} \, dS = 3 \iiint dv = 3 \times \text{vol.}$$

Or in other words, three times the volume included by any closed surface is obtained by multiplying every element of surface by the perpendicular from the origin to its plane and adding the results.

In a sphere, for instance, taking the origin at the center, $\mathbf{r}$ is perpendicular to every element of surface and is of constant length, therefore $\mathbf{n} \cdot \mathbf{r} = r$,

$$3 \times \text{vol sphere} = r \iint dS = 4 \pi r^3,$$

a true result.

To obtain $\nabla \cdot \mathbf{r}_1$. By (123) and (128),

$$\nabla \cdot \mathbf{r} = 3 = \nabla \cdot (r \mathbf{r}_1) = r \nabla \cdot \mathbf{r}_1 + \mathbf{r}_1 \cdot \nabla r = r \nabla \cdot \mathbf{r}_1 + \frac{dr}{dr} \quad \text{(by (112))}$$

$$= r \nabla \cdot \mathbf{r}_1 + 1.$$

Hence $$\nabla \cdot \mathbf{r}_1 = \frac{2}{r}.$$

Equation of the Flow of Heat. As another example of the use of the Divergence Theorem (121) consider the general laws of thermal flow. Consider a volume of matter through which heat is flowing, and consider a surface S drawn anywhere in this space. Let $\mathbf{q}$ be the flux of heat, or in other words, the amount of heat which crosses unit area drawn normally to the lines of flow per unit time; $\mathbf{q}$ is also called the heat current-density.

The amount of heat which escapes through the surface in any time is furnished at the expense of the material inside that surface which must then be cooling off at a certain rate.

By Fourier's Law, § 47, the heat flows in the direction of greatest decrease in temperature, θ, and with an intensity proportional to a property of the material through which it is flowing, called its heat or thermal conductivity k. So that,

$$\mathbf{q} = - k\nabla\theta.$$

The coefficient k may vary from point to point of the medium, and may be also a function of the temperature. In most practical applications it is assumed to be constant.

If there are no sources nor sinks of heat within the surface any elementary volume dv is cooling at some rate $-\frac{\partial\theta}{\partial t}$. The amount of heat which leaves this elementary volume in unit time must then be, if ρ be its density and c its specific heat,

$$-\frac{\partial\theta}{\partial t} c\,\rho\, dv.$$

For the whole volume, S, the heat lost, which must be equal to that passing through the surface, is

$$\iiint_{\text{vol}} -\frac{\partial\theta}{\partial t} c\,\rho\, dv = \iint_{s} \mathbf{n}\cdot\mathbf{q}\, dS.$$

By the Divergence Theorem the surface integral is

$$\iiint_{\text{vol}} \nabla\cdot\mathbf{q}\, dv,$$

and since $\quad \mathbf{q} = - k\nabla\theta, \qquad \nabla\cdot\mathbf{q} = -\nabla\cdot k\nabla\theta.$
So that

$$\iiint_{\text{vol}} -\frac{\partial\theta}{\partial t} c\,\rho\, dv = \iiint_{\text{vol}} -\nabla\cdot k\,\nabla\theta\, dv.$$

Since this equation holds whatever surface be considered, the integrands are equal everywhere and

$$\frac{\partial\theta}{\partial t} c\rho = \nabla\cdot k\,\nabla\theta.$$

If k be assumed constant this becomes

$$\frac{\partial \theta}{\partial t} = \frac{k}{c\rho} \nabla^2 \theta$$

$$= a^2 \nabla^2 \theta, \text{ where } a^2 = \frac{k}{c\rho}. \qquad (107)$$

This is the general differential equation for the flow of heat in a body.

If the steady state is reached, that is, if the temperatures are everywhere constant (this does not mean the *same* everywhere), it becomes

$$\nabla^2 \theta = 0,$$

independently of the values of k, ρ, and c; that is, the distribution of temperature follows the same law as the distribution of potential according to Laplace's Equation (157). So that, what is true about the potential, is under analogous conditions true of temperature, and the two subjects, temperature distribution and potential, become *identical* in mathematical treatment.

53. Equation of Continuity. Considering again a moving liquid, if there are no sources nor sinks of the fluid in the region considered, then the equation

$$\nabla \cdot \mathbf{q} = \text{div } \mathbf{q} = 0 \qquad (124)$$

expresses the condition that the fluid does not concentrate towards nor expand from any point, as this is the only remaining way by which more liquid can leave any small closed surface than can enter it, or conversely. In other words, it means incompressibility. This equation is called the equation of continuity. It is of great importance in electricity, as according to the theory of Maxwell the electric displacement behaves like an incompressible fluid. If the divergence does exist it means that at the point considered there must be a source of lines of force or what is the same thing, electricity.

Solenoidal Distribution of a Vector. Should the divergence of a vector function be zero everywhere, then always as much vector flux enters *any* volume element as leaves it, or in other words, the lines of vector flux cannot end nor begin in free space. They must then form closed curves or end at infinity. Such a vector distribution is called *solenoidal.* For example, the motion of any incompressible fluid such as water, gives a velocity distribution which is solenoidal.

54. Curl. The Operator $\nabla\times$ applied to $\mathbf{F}$ or curl $\mathbf{F}$ (read del cross $\mathbf{F}$ or curl of $\mathbf{F}$), also sometimes written in German books, rot $\mathbf{F}$ (read rotation of $\mathbf{F}$), is a new *vector* derived from $\mathbf{F}$. Like ∇ it is invariant to choice of axes and has an important significance in physics. It may be defined by the equation

$$\nabla\times\mathbf{F} \equiv \text{curl } \mathbf{F} \equiv \begin{vmatrix} \mathbf{i} & \mathbf{j} & \mathbf{k} \\ \dfrac{\partial}{\partial x} & \dfrac{\partial}{\partial y} & \dfrac{\partial}{\partial z} \\ F_1 & F_2 & F_3 \end{vmatrix} = \mathbf{i}\left(\frac{\partial F_3}{\partial y} - \frac{\partial F_2}{\partial z}\right) + \mathbf{j}\left(\frac{\partial F_1}{\partial z} - \frac{\partial F_3}{\partial x}\right) + \mathbf{k}\left(\frac{\partial F_2}{\partial x} - \frac{\partial F_1}{\partial y}\right). \quad (125)$$

The new vector, curl $\mathbf{F}$, has components

$$\left(\frac{\partial F_3}{\partial y} - \frac{\partial F_2}{\partial z}\right), \quad \left(\frac{\partial F_1}{\partial z} - \frac{\partial F_3}{\partial x}\right), \quad \left(\frac{\partial F_2}{\partial x} - \frac{\partial F_1}{\partial y}\right),$$

along the three axes. When applied to a vector function $\nabla\times$ gives a result independent of the axes because ∇ itself is independent of them. **We may say in general that all combinations of ∇ with vector or scalar point-functions give results independent of any choice of axes.** By a direct transformation from one set of axes to another we may prove the invariant property of these operators and thus eliminate any lingering doubt in the mind of the skeptic. To a physicist, however, to say that the operations of ∇ upon any functions are dependent upon a choice of axes, is like saying that the physical properties of any medium depend upon the language in which you express them. For instance, we have shown (§ 47) that ∇V, where V is the potential say, gives rise to a vector showing the direction in which V changes most rapidly and its magnitude.

What have *axes* to do with such a result? It is true whatever kind of coördinates are used, however placed, or even if none are used at all.

We are here dealing with the properties themselves, and not with any particular method of representing them. It is in this respect that the analysis of vectors is extremely useful, as by its intelligent study clear conceptions must necessarily be obtained.

Example of Curl. In order to give an idea of the meaning of the $\nabla\times$ or curl of a vector function, consider the general motion of a rigid body. We have seen (§ 22) that the motion may be resolved into a velocity of translation $\mathbf{q}_0$ of the origin chosen arbitrarily and an angular velocity of rotation $\boldsymbol{\omega}$ about a line passing through this origin. The velocity $\mathbf{q}$ of any point $\mathbf{r}$ is then given by

$$\mathbf{q} = \mathbf{q}_0 + \boldsymbol{\omega}\times\mathbf{r},$$

where $\mathbf{q}_0$ and $\boldsymbol{\omega}$ are the same for all points in the body at any given instant. Taking the curl of this equation, or, in other words, applying or operating with $\nabla\times$, we obtain

$$\nabla\times\mathbf{q} = \nabla\times\mathbf{q}_0 + \nabla\times(\boldsymbol{\omega}\times\mathbf{r}).$$

Since $\mathbf{q}_0$ is a constant throughout the body $\nabla\times\mathbf{q}_0 = 0$ and

$$\nabla\times\mathbf{q} = \nabla\times(\boldsymbol{\omega}\times\mathbf{r}).$$

In this product ∇ differentiates $\mathbf{r}$ alone because $\boldsymbol{\omega}$ is a constant throughout the body at any time. In order to find the value of this expression expand the triple vector product by (55), considering ∇ as an ordinary vector,

$$\nabla\times(\boldsymbol{\omega}\times\mathbf{r}) = \boldsymbol{\omega}(\nabla_r\cdot\mathbf{r}) - \mathbf{r}(\nabla_r\cdot\boldsymbol{\omega}).$$

As ∇ cannot act upon $\boldsymbol{\omega}$, we interpret the last term as $(\boldsymbol{\omega}\cdot\nabla_r)\mathbf{r}$,

but $\quad \nabla\cdot\mathbf{r} = 3$ by (123) and $(\boldsymbol{\omega}\cdot\nabla)\,\mathbf{r} = \boldsymbol{\omega}$ by (118),

$$\nabla\times\mathbf{q} = 3\,\boldsymbol{\omega} - \boldsymbol{\omega} = 2\,\boldsymbol{\omega},$$

and

$$\boldsymbol{\omega} = \tfrac{1}{2}\,\nabla\times\mathbf{q} = \tfrac{1}{2}\,\text{curl}\,\mathbf{q}. \tag{126}$$

See also equation 131.

We see then that when a rigid system is in motion the operator $\nabla\times$ applied to its velocity-function gives twice its angular velocity in magnitude and direction. We may then write

$$\mathbf{q} = \mathbf{q}_0 + \tfrac{1}{2}\,\text{curl}\,\mathbf{q}\times\mathbf{r}.$$

Consider now a very small portion of a fluid such that the portion may be considered to move as a rigid body for the instant; it is fairly evident that the curl of the velocity there would give similarly twice its angular velocity of rotation. The curl or $\nabla\times$ is an operator such that when applied to any velocity-function it gives twice the angular velocity of rotation at any point in direction and magnitude.

55. Motion of Rotation which has No Curl. Irrotational Motion. A clearer idea of curl may perhaps be given by a consideration of the following two possible motions of a fluid about an axis. Considering Fig. 53, if the infinitesimal portions of the fluid, indicated by short straight lines, move from position 1 to position 2, as indicated, then evidently every elementary portion of the fluid has rotated by the same amount and the operator $\nabla\times$ would give this rotation multiplied by 2. On the contrary, if the infinitesimal elements in moving about the axis O do not rotate but remain facing one way as in *B*, then the curl of such a motion would be zero.* Superficially, however, to the eye the two motions here described would look the same. If we assume that the molecules of iron are free to rotate we may realize these two motions. If a piece of iron were rotated in a strong magnetic field, the molecules constantly pointing in the fixed direction of magnetic induction, we should obtain a motion such as B,

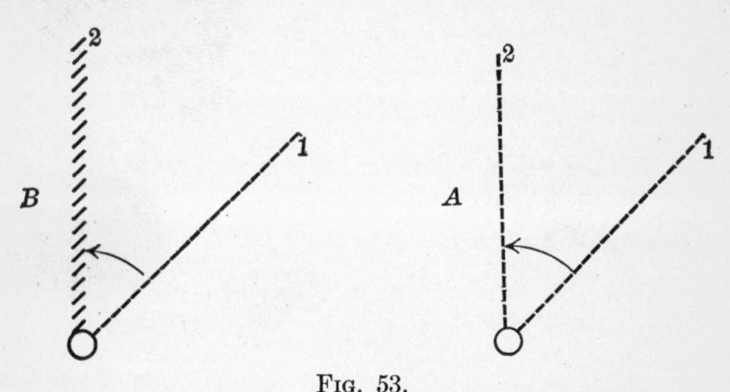

FIG. 53.

while if rotated in a non-magnetic field the motion would be similar to A.

Any motion which has a curl is said to be *rotational* or *vortical;* if it has no curl it is called *irrotational* or *non-vortical.* Any motion represented by a function whose curl is zero is one in which the infinitesimal elements do not rotate, and conversely.

56. ∇, $\nabla\cdot$ and $\nabla\times$ Applied to Various Functions. It is frequently necessary in many cases to apply the operators formed with ∇ to combinations of scalar and vector functions. The following rules will be found useful for reference.

* A rigid body whose elementary parts move with it as in A, Fig. 53, might be called *atomically-rigid;* a rigid body whose elementary parts move as in B would then be *non-atomically-rigid.* A calculation was made to see whether the difference in the moment of inertia of these two kinds of motion could be observed in the case of iron. But molecules are so small that the calculated difference in moment of inertia could not be observed by the most sensitive laboratory methods.

Let u and v be scalar point-functions; **u** and **v** vector point-functions.

Then

$$\begin{aligned} \nabla(u+v) &= \nabla u + \nabla v, \\ \nabla\cdot(\mathbf{u}+\mathbf{v}) &= \nabla\cdot\mathbf{u} + \nabla\cdot\mathbf{v}, \\ \nabla\times(\mathbf{u}+\mathbf{v}) &= \nabla\times\mathbf{u} + \nabla\times\mathbf{v}. \end{aligned} \tag{127}$$

$$\begin{aligned} \nabla(uv) &= v\nabla u + u\nabla v \\ \nabla\cdot(u\,\mathbf{v}) &= \nabla u\cdot\mathbf{v} + u\,\nabla\cdot\mathbf{v} \\ \nabla\times(u\,\mathbf{v}) &= \nabla u\times\mathbf{v} + u\,\nabla\times\mathbf{v}. \end{aligned} \tag{128}$$

$$\begin{aligned} \nabla(\mathbf{u}\cdot\mathbf{v}) &= \mathbf{u}\cdot\nabla\mathbf{v} + \mathbf{u}\times(\nabla\times\mathbf{v}) + \mathbf{v}\cdot\nabla\mathbf{u} + \mathbf{v}\times(\nabla\times\mathbf{u}) \\ &= \nabla_v(\mathbf{u}\cdot\mathbf{v}) + \nabla_u(\mathbf{u}\cdot\mathbf{v}). \end{aligned} \tag{129}$$

$$\nabla\cdot(\mathbf{u}\times\mathbf{v}) = \mathbf{v}\cdot\nabla\times\mathbf{u} - \mathbf{u}\cdot\nabla\times\mathbf{v}. \tag{130}$$

$$\begin{aligned} \nabla\times(\mathbf{u}\times\mathbf{v}) &= \mathbf{u}\,(\nabla_{uv}\cdot\mathbf{v}) - \mathbf{v}\,(\nabla_{uv}\cdot\mathbf{u}) \\ &= \mathbf{u}\,\nabla_v\cdot\mathbf{v} + \mathbf{v}\cdot\nabla_u\mathbf{u} - \mathbf{v}\nabla_u\cdot\mathbf{u} - \mathbf{u}\cdot\nabla_v\mathbf{v}. \end{aligned} \tag{131}$$

The convention here used, is that the operator ∇ applies to the nearest term when there are no parentheses or else the variables on which it *operates* are indicated by subscripts to it. So that, for example,

$$\nabla uv \text{ means } (\nabla u)\,v \text{ and not } \nabla(uv),$$

and $\nabla u\cdot\mathbf{v}$ means $(\nabla u)\cdot\mathbf{v}$. In this last case it could mean nothing else, as in $\nabla(u\cdot\mathbf{v})$, $u\cdot\mathbf{v}$ being a scalar product of a *scalar* and a vector can have no meaning.

In the above fundamental formulæ subscripts have been used to render ambiguity impossible.

Methods of Proof of the Formulæ. These formulæ may *all* be verified by expanding the quantities in terms of their components along **i j k**, differentiating, and rearranging.

If we remember that the symbolic components of ∇, $\frac{\partial}{\partial x}$, $\frac{\partial}{\partial y}$, $\frac{\partial}{\partial z}$ obey with the components of any other vector or of any other ∇ all the laws of common algebra, we should expect ∇ to obey the same laws as any other vector in combination with vectors or other ∇'s. With this in mind the majority of the formulæ on page 121 can be written down at once without relying upon the demonstration outlined above.

It is evident at once from the definition of ∇ that

$$\nabla(u + v) = \nabla u + \nabla v.$$

Take the formula for instance

$$\nabla\cdot(u\,\mathbf{v}) = \nabla u\cdot\mathbf{v} + u\nabla\cdot\mathbf{v}.$$

The ∇ is supposed to operate upon both u and $\mathbf{v}$. All the possible combinations of ∇ and a dot with u and $\mathbf{v}$ are formed which can have a meaning, letting ∇ act once upon each variable. In the above example, as u is a scalar, ∇ can act on it only as ∇u. The dot, which is as yet unemployed, is used in forming the scalar product of ∇u, a vector, with $\mathbf{v}$, another vector. As yet ∇ has not differentiated $\mathbf{v}$. In the second term the only way it can act on $\mathbf{v}$ is by forming a scalar product, giving $\nabla\cdot\mathbf{v}$, which multiplied by u is $u\nabla\cdot\mathbf{v}$, the correct result.

As another example, consider the expansion for $\nabla\times(\mathbf{u}\times\mathbf{v})$. We expand this triple vector product as usual, considering ∇ to be an ordinary vector,

$$\nabla\times(\mathbf{u}\times\mathbf{v}) = \mathbf{u}\,(\nabla_{uv}\cdot\mathbf{v}) - \mathbf{v}\,(\nabla_{uv}\cdot\mathbf{u}),$$

where be it remembered that ∇ is to differentiate both $\mathbf{u}$ and $\mathbf{v}$ in *each* of the terms. This is here specifically indicated by the use of subscripts. From $\mathbf{u}\,(\nabla_{uv}\cdot\mathbf{v})$ can be formed $\mathbf{u}(\nabla_v\cdot\mathbf{v})$ and $(\mathbf{v}\cdot\nabla_u)\mathbf{u}$ only; so that

$$\mathbf{u}(\nabla_{uv}\cdot\mathbf{v}) = \mathbf{u}\,(\nabla_v\cdot\mathbf{v}) + (\mathbf{v}\cdot\nabla_u)\,\mathbf{u}$$

similarly from $\mathbf{v}\,(\nabla_{uv}\cdot\mathbf{u})$, $\mathbf{v}\,(\nabla_u\cdot\mathbf{u})$ and $(\mathbf{u}\cdot\nabla_v)\,\mathbf{v}$ can be derived, so that

$$\mathbf{v}\,(\nabla_{uv}\cdot\mathbf{u}) = \mathbf{v}\,(\nabla_u\cdot\mathbf{u}) + (\mathbf{u}\cdot\nabla_v)\,\mathbf{v}.$$

The single subscripts as here used are not necessary according to the convention explained above. We then have

$$\nabla\times(\mathbf{u}\times\mathbf{v}) = \mathbf{u}\nabla\cdot\mathbf{v} + \mathbf{v}\cdot\nabla\mathbf{u} - \mathbf{v}\nabla\cdot\mathbf{u} - \mathbf{u}\cdot\nabla\mathbf{v}, \qquad (132)$$

where the ∇ on the left is to operate on both **u** and **v**, while on the right it operates only on the vector following it. This kind of notation is exactly similar to

$$\frac{d}{dx}(uv) = \left(\frac{d}{dx}u\right)v + u\left(\frac{d}{dx}v\right).$$
$$= \frac{du}{dx}v + u\frac{dv}{dx}.$$

Consider the expression $\mathbf{v}\times(\nabla_u\times\mathbf{u})$ in which ∇ is to act upon **u** alone. Expanding,

$$\mathbf{v}\times(\nabla_u\times\mathbf{u}) = \nabla_u(\mathbf{u}\cdot\mathbf{v}) - \mathbf{u}(\nabla_u\cdot\mathbf{v})$$
$$= \nabla_u(\mathbf{u}\cdot\mathbf{v}) - (\mathbf{v}\cdot\nabla_u)\,\mathbf{u}.$$

Similarly

$$\mathbf{u}\times(\nabla_v\times\mathbf{v}) = \nabla_v(\mathbf{u}\cdot\mathbf{v}) - (\mathbf{u}\cdot\nabla_v)\,\mathbf{v}.$$

Adding the two equations, we combine $\nabla_u(\mathbf{u}\cdot\mathbf{v}) + \nabla_v(\mathbf{u}\cdot\mathbf{v})$ into $\nabla(\mathbf{u}\cdot\mathbf{v})$ by definition, hence equation (129).

The notation

$$\nabla(\mathbf{u}\cdot\mathbf{v}) = \nabla_u(\mathbf{u}\cdot\mathbf{v}) + \nabla_v(\mathbf{u}\cdot\mathbf{v})$$

is strictly analogous to

$$d(\mathbf{u}\cdot\mathbf{v}) = d_u(\mathbf{u}\cdot\mathbf{v}) + d_v(\mathbf{u}\cdot\mathbf{v}),$$

which corresponds to *partial* differentiation, and is true for the same reasons.

We may write also

$$\nabla\times(u\,\mathbf{v}) = \nabla_u\times(u\,\mathbf{v}) + \nabla_v\times(u\,\mathbf{v}),$$

or

$$\nabla\times(\mathbf{u}\times\mathbf{v}) = \nabla_u\times(\mathbf{u}\times\mathbf{v}) + \nabla_v\times(\mathbf{u}\times\mathbf{v}), \text{ etc.}$$

The process outlined above will always lead to correct results. It is something more than a help to the memory. A general rigid mathematical proof of its validity has been given.*

* See to this effect Joly. Manual of Quaternions, p. 75.

57. Expansion Analogous to Taylor's Theorem. Expanding as a triple vector product and assuming that ∇ acts on v alone, we have

$$\mathbf{u}\times(\nabla\times\mathbf{v}) = \nabla_v(\mathbf{u}\cdot\mathbf{v}) - \mathbf{v}(\nabla_v\cdot\mathbf{u})$$

and

$$= \nabla_v(\mathbf{u}\cdot\mathbf{v}) - \mathbf{u}\cdot\nabla\mathbf{v},$$

or

$$\mathbf{u}\cdot\nabla\mathbf{v} = \nabla_v(\mathbf{u}\cdot\mathbf{v}) - \mathbf{u}\times(\nabla\times\mathbf{v}). \tag{133}$$

If $\mathbf{u} = \mathbf{r}_1$, a unit vector, and $\mathbf{v} = \mathbf{q}$, then (133) becomes

$$\mathbf{r}_1\cdot\nabla\mathbf{q} = \nabla_q(\mathbf{r}_1\cdot\mathbf{q}) - \mathbf{r}_1\times \text{curl } \mathbf{q}, \tag{134}$$

and states that the directional derivative of a vector function $\mathbf{q}$ in the direction $\mathbf{r}_1$ is equal to the derivative of the projection of $\mathbf{q}$ in that direction plus the vector product of the curl of $\mathbf{q}$ into that direction.

Multiplying the directional derivative by dr_0, we obtain the difference in $\mathbf{q}$ due to a displacement $d\mathbf{r}$ in the direction $\mathbf{r}_1$; this gives then, if $\mathbf{q}_r$ is the value of $\mathbf{q}$ at the end of $\mathbf{r}$ and $\mathbf{q}_{r+dr}$ is the value of $\mathbf{q}$ at the end of $\mathbf{r} + d\mathbf{r}$,

$$d\mathbf{q} = d\mathbf{r}\cdot\nabla\mathbf{q} = \mathbf{q}_{r+dr} - \mathbf{q}_r = \nabla_q(d\mathbf{r}\cdot\mathbf{q}) - d\mathbf{r}\times(\nabla_q\times\mathbf{q}).$$

So that

$$\mathbf{q}_{r+dr} = \mathbf{q}_r + \nabla_q(d\mathbf{r}\cdot\mathbf{q}) + (\nabla\times\mathbf{q})\times d\mathbf{r}, \tag{135}$$

or

$$\mathbf{q}(\mathbf{r} + d\mathbf{r}) = \mathbf{q}(\mathbf{r}) + \nabla_q[d\mathbf{r}\cdot\mathbf{q}(\mathbf{r})] + [\nabla\times\mathbf{q}(\mathbf{r})]\times d\mathbf{r}.$$

This equation is analogous to the expansion of a function by means of Taylor's theorem.

58. Theorem Due to Stokes. The line-integral of a vector function F around any closed contour is equal to the surface integral of the curl of that function over any surface of which the contour is a bounding edge.

In symbols this is

$$\int_{\mathfrak{S}} \mathbf{F}\cdot d\mathbf{r} = \iint_{\text{Cap}} \mathbf{n}\cdot\text{curl } \mathbf{F}\, dS. \tag{136}$$

This important theorem may be demonstrated in a number of different ways. The following is a demonstration given by Helmholtz depending upon the *variation* of a line-integral. The principle of commutativity of δ with d and $\int$ is all that is needed to assume here. Consider the line-integral J of the vector point-function $\mathbf{F}\ (\mathbf{r})$ along the path ACB.

$$J = \int_B^A \mathbf{F} \cdot d\mathbf{r} \quad \text{(path } ACB\text{)}.$$

The possibility of computing this integral in general depends entirely on the path ACB, and with this under-

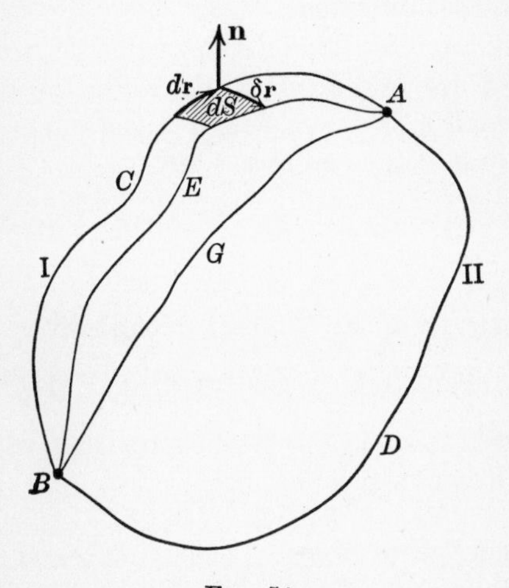

FIG. 54.

standing it is perfectly definite. It is now required to find the variation in this integral when the path ACB is varied into an adjacent one AEB infinitesimally close to ACB but differing from it in an arbitrary manner. The two paths, however, are to begin at A and end at B, two fixed points on the contour. Taking the variation of the integral we obtain

$$\delta J = \delta \int_A^B \mathbf{F}\cdot d\mathbf{r} = \int_A^B \delta\,(\mathbf{F}\cdot d\mathbf{r}) = \int_A^B \delta\,\mathbf{F}\cdot d\mathbf{r} + \int_A^B \mathbf{F}\cdot\delta d\mathbf{r}.$$

This becomes, by an interchange of d and δ in the last integral and with an integration by parts,

$$\delta J = \mathbf{F}\cdot\delta\mathbf{r}\Big/_A^B + \int_A^B (\delta\,\mathbf{F}\cdot d\mathbf{r} - d\,\mathbf{F}\cdot\delta\mathbf{r}).$$

The integrated term is zero, for since the limits A and B are fixed, there can be no variation $\delta\,\mathbf{r}$ at these points. Remembering also (§ 50) that

$$\delta\,\mathbf{F} = \delta\mathbf{r}\cdot\nabla_F\mathbf{F} \quad\text{and}\quad d\,\mathbf{F} = d\mathbf{r}\cdot\nabla_F\mathbf{F},$$

$$\delta J = \int_A^B d\mathbf{r}\cdot(\delta\mathbf{r}\cdot\nabla_F)\,\mathbf{F} - \delta\mathbf{r}\cdot(d\mathbf{r}\cdot\nabla_F)\,\mathbf{F}$$

$$= \int_A^B \delta\mathbf{r}\cdot\nabla_F(d\mathbf{r}\cdot\mathbf{F}) - d\,\mathbf{r}\cdot\nabla_F(\delta\mathbf{r}\cdot\mathbf{F}), \text{ using (117)}$$

where ∇ acts on $\mathbf{F}$ alone.

By (58) this may be written as

$$\delta J = \int_A^B (\delta\mathbf{r}\times d\mathbf{r})\cdot(\nabla\times\mathbf{F}).$$

Or more directly by (129),

$$\delta\mathbf{r}\cdot\nabla\mathbf{F} = \nabla_F(\delta\mathbf{r}\cdot\mathbf{F}) - \delta\mathbf{r}\times(\nabla\times\mathbf{F}),$$

hence

$$d\mathbf{r}\cdot(\delta\mathbf{r}\cdot\nabla\mathbf{F}) = d\mathbf{r}\cdot\nabla_F(\delta\mathbf{r}\cdot\mathbf{F}) - d\mathbf{r}\cdot\delta\mathbf{r}\times(\nabla\times\mathbf{F}).$$

But by (117) the left side is $\delta\mathbf{r}\cdot\nabla_F\,(d\mathbf{r}\cdot\mathbf{F})$.
Hence

$$\delta\mathbf{r}\cdot\nabla_F\,(d\mathbf{r}\cdot\mathbf{F}) - d\mathbf{r}\cdot\nabla_F\,(\delta\mathbf{r}\cdot\mathbf{F}) = \delta\mathbf{r}\times d\mathbf{r}\cdot\nabla\times\mathbf{F}.$$

Referring to the figure it is seen that $\delta\mathbf{r}\times d\mathbf{r}$ is the vector area of the infinitesimal surface formed by $\delta\mathbf{r}$ and $d\mathbf{r}$, so that calling **n** the *unit* normal to the elementary area dS, we may write $\delta\mathbf{r}\times d\mathbf{r} = \mathbf{n}\, dS$, so that

$$\delta J = \int_A^B \mathbf{n}\cdot(\nabla\times\mathbf{F})\, dS.$$

Another infinitesimal transformation is now made to a new curve AGB having the same fixed ends A and B and so on, until the movable path has swept over the surface included between the limiting curves I and II. Let now the sum of all the variations in J be added together. The result will be equal to the difference between J_1 and J_2, the values of J for the extreme paths; hence

$$J_2 - J_1 = \lim_{\delta \mathbf{r} \doteq 0} \sum \delta J = \int\int \mathbf{n}\cdot(\nabla\times\mathbf{F})\, dS.$$

But $-J_1$ is the line-integral from A to B along BCA, and therefore $J_2 - J_1$ is the value of the line-integral *around* the contour $ADBCA$. So that

$$\int_{\circlearrowleft} \mathbf{F}\cdot d\mathbf{r} = \int\int_{\text{cap}} (\mathbf{n}\cdot\text{curl}\ \mathbf{F})\, dS. \qquad (136)$$

This is **Stokes' Theorem.***

59. Condition for the Vanishing of the Curl of a Vector-Function. In the above demonstration, if the value of the integral J is the same whatever path is taken between A and B, and if this is true wherever the points A and B are taken, it follows that J_1 always equals J_2, so that for *any* surface S

$$\int\int_S \mathbf{n}\cdot\text{curl}\ \mathbf{F}\, dS = 0 \quad \text{and hence} \quad \nabla\times\mathbf{F} = \text{curl}\ \mathbf{F} = 0 \qquad (137)$$

must be true everywhere. In this case the value of I depends only upon the position of the ends of the path and in no wise

* See also for other demonstrations of Stokes' Theorem,
Bucherer, Elemente der Vektor-Analysis, pp. 42–44.
Gibbs-Wilson, Vector Analysis, pp. 188–190.
Gans: Einführung in die Vektoranalysis, pp. 35–39.
See Appendix, p. 249, for two other proofs.

upon the shape of it. Conversely, if the curl $\mathbf{F} = 0$ in a region, then the line-integral of $\mathbf{F}$ between any two points A and B in the region is independent of the path chosen between them. In this case if one end A of a curve is fixed, the value of the integral $\int_A^B \mathbf{F} \cdot d\mathbf{r}$ is simply a function of its upper limit B whatever the path from A to B may be. Let ϕ denote this *scalar* function. The integrand $\mathbf{F} \cdot d\mathbf{r}$ must then be a perfect differential and hence of the form $d\phi$, which by (114) is the same as

$$d\phi = d\,\mathbf{r} \cdot \nabla \phi,$$

so that for all values of $d\mathbf{r}$,

$$d\mathbf{r} \cdot \nabla \phi = d\mathbf{r} \cdot \mathbf{F}$$

and hence

$$\mathbf{F} = \nabla \phi. \tag{138}$$

Or in other words, if $\mathbf{F}$ has no curl, it is the rate of fastest increase $\nabla \phi$ or grad ϕ of a scalar function

$$\phi_B = \int_A^B \mathbf{F} \cdot d\mathbf{r} + \phi_A,$$

where ϕ_A is a constant.

The scalar function ϕ thus determined is called the potential of $\mathbf{F}$. As we have seen before, a scalar function divides space up into shells or laminæ by means of its level surfaces. The vector-function $\mathbf{F} = \nabla \phi$ derived from such a function is for this reason said to be a **lamellar** vector (Maxwell). The curl of a lamellar vector is then always zero,

or

$$\text{curl}\,(\nabla \phi) = 0. \tag{139}$$

Conservative System of Forces. By Stokes' Theorem (136) we see that if the line-integral of $\mathbf{F} \cdot d\mathbf{r}$, *i.e.*, the work around any closed path, always vanishes, then the forces have no curl; and also that in this case the forces in the field are derivable from a potential function ϕ. Such a system of forces is a *Conservative System*, and we may define such a system:

When the forces acting on a system of bodies are of such a nature that the algebraic total of the work done in performing any series of displacements which bring the system back to its original configuration is nil, the system of forces is said to be conservative. The condition, then, for a system of forces $\mathbf{F}$ to be conservative is

$$\text{curl } \mathbf{F} \equiv \nabla\times\mathbf{F} = 0.$$

60. Condition for a Perfect Differential. If the total differential $d\phi$ of a scalar point-function $\phi(\mathbf{r}) = C$ be taken, we obtain, by (§ 49), the equation

$$d\phi = d\mathbf{r}\cdot\nabla\phi = 0.$$

This is of the form $$d\mathbf{r}\cdot\mathbf{f}(\mathbf{r}) = 0,$$

which in general is *not* a perfect differential. In Cartesian notation this equation becomes, if f_1, f_2, f_3 are the components of $\mathbf{f}$ along the three axes,

$$f_1dx + f_2dy + f_3dz = 0,$$

and our problem is to find the condition for integrability of it. Assume that, by multiplying this differential equation by some scalar factor μ, it may be made a perfect differential, or that for every value of $d\mathbf{r}$

$$\mu(d\mathbf{r}\cdot\mathbf{f}) = d\mathbf{r}\cdot\mu\,\mathbf{f} \equiv d\mathbf{r}\cdot\nabla\phi.$$

μ is called an **integrating factor.** In this case, then,

$$\mu\,\mathbf{f} = \nabla\phi.$$

In order to eliminate ϕ take the curl of this equation, because we know that the curl of a lamellar vector (139) is zero,

$$\nabla\times(\mu\,\mathbf{f}) = 0,$$

which may be expanded as

$$\nabla\times(\mu\,\mathbf{f}) = \mu\nabla\times\mathbf{f} + (\nabla\mu)\times\mathbf{f} = 0.$$

Applying $\mathbf{f}\cdot$ to eliminate the second term, because $\mathbf{a}\cdot\mathbf{a}\times\mathbf{b} \equiv 0$ in general, there remains

$$\mathbf{f}\cdot(\mu\nabla\times\mathbf{f}) = \mu\,(\mathbf{f}\cdot\nabla\times\mathbf{f}) = 0.$$

So that finally we find that the condition that the equation defined by $\mathbf{f}\cdot d\mathbf{r} = 0$ should be integrable is that $\mathbf{f}$ and its curl, $\nabla\times\mathbf{f}$, shall be at right angles or that the curl vanish. To put this discussion in a more familiar form, we have proved that the condition of integrability of

$$\mathbf{F}\cdot d\mathbf{r} = X dx + Y dy + Z dz$$

is that

$$\mathbf{F}\cdot\nabla\times\mathbf{F} = \mathbf{0} \tag{140}$$

or $$X\left(\frac{\partial Z}{\partial y} - \frac{\partial Y}{\partial z}\right) + Y\left(\frac{\partial X}{\partial z} - \frac{\partial Z}{\partial x}\right) + Z\left(\frac{\partial Y}{\partial x} - \frac{\partial X}{\partial y}\right) = 0,$$

a well-known result.

If the

$$\text{curl}\ \mathbf{F} = 0,$$

this equation is evidently satisfied.

If the curl $\mathbf{F}$ is not zero, the equation says that it must be everywhere perpendicular to $\mathbf{F}$.

For example,

$$yz\,dx + zx\,dy + xy\,dz$$

has no curl and is therefore derivable from a single primitive, *i.e.*,

$$xyz = \text{const.}$$

On the other hand,

$$ay^2z^2dx + bz^2x^2dy + cx^2y^2dz$$

has a curl, but this curl,

$$2\,x^2(cy - bz)\,\mathbf{i} + 2\,y^2(az - cx)\,\mathbf{j} + 2\,z^2(bx - ay)\,\mathbf{k},$$

and the vector $\mathbf{F}$,

$$ay^2z^2\mathbf{i} + bz^2x^2\mathbf{j} + cx^2y^2\mathbf{k},$$

are at right angles, and this equation is also derivable from a single primitive, *i.e.*,

$$\frac{a}{x} + \frac{b}{y} + \frac{c}{z} = \text{const.}$$

61. Taylor's Theorem. The Operator $e^{\epsilon\cdot\nabla}(\quad)$. Taylor's Theorem is often written concisely as

$$\begin{aligned} f(x+h, y+k, z+l) = f(xyz) &+ \frac{1}{1!}\left(h\frac{\partial}{\partial x} + k\frac{\partial}{\partial y} + l\frac{\partial}{\partial z}\right) f(xyz) \\ &+ \frac{1}{2!}\left(h\frac{\partial}{\partial x} + k\frac{\partial}{\partial y} + l\frac{\partial}{\partial z}\right)^2 f(xyz) \\ &+ \frac{1}{3!}\left(h\frac{\partial}{\partial x} + k\frac{\partial}{\partial y} + l\frac{\partial}{\partial z}\right)^3 f(xyz) + \ldots \end{aligned}$$

If the components of $\mathbf{r}$ be x, y, z, and those of ϵ be h, k, l, it may still further be condensed into

$$f(\mathbf{r}+\epsilon) = f(\mathbf{r}) + \epsilon\cdot\nabla f(\mathbf{r}) + \frac{1}{2!}(\epsilon\cdot\nabla)^2 f(\mathbf{r}) + \frac{1}{3!}(\epsilon\cdot\nabla)^3 f(\mathbf{r}) + \cdots$$

Remembering the expansion for e^x,

$$e^x = 1 + \frac{x}{1!} + \frac{x^2}{2!} + \frac{x^3}{3!} + \cdots,$$

we may write the last equation symbolically in the still shorter form,

$$f(\mathbf{r} + \epsilon) = e^{\epsilon\cdot\nabla} f(\mathbf{r}), \tag{141}$$

so that the symbolic differential operator $e^{\epsilon\cdot\nabla}$ acting on any function $f(\mathbf{r})$ gives its value when $\mathbf{r}$ becomes $\mathbf{r} + \epsilon$.

62. Euler's Theorem on Homogeneous Functions. We may employ this equation to demonstrate a useful theorem due to Euler and known by his name. A function ϕ of degree n in a variable $\mathbf{r}$ is said to be homogeneous when $\mathbf{r}$ occurs the same number of times in every term of it. It is one such that

$$\phi(a\,\mathbf{r}) = a^n \phi(\mathbf{r}), \tag{142}$$

where a is any constant. Apply Taylor's Theorem to the homogeneous function $\phi(\mathbf{r})$ and let $\boldsymbol{\epsilon} = g\,\mathbf{r}$, where g is a small scalar multiplier. Then

$$e^{g\,\mathbf{r}\cdot\nabla}\phi(\mathbf{r}) = \phi(\mathbf{r} + g\,\mathbf{r}) = \phi[\mathbf{r}(1+g)] = (1+g)^n\phi(\mathbf{r}),$$

and hence

$$\phi(\mathbf{r}) + g\,\mathbf{r}\cdot\nabla\phi(\mathbf{r}) + \frac{g^2}{1\cdot 2}(\mathbf{r}\cdot\nabla)^2\phi(\mathbf{r}) + \cdots$$
$$= \left[1 + ng + \frac{n(n-1)}{1\cdot 2}g^2 + \cdots\right]\phi(\mathbf{r}).$$

Subtracting $\phi(\mathbf{r})$ from both sides and dividing through by g there remains

$$\mathbf{r}\cdot\nabla\phi(\mathbf{r}) + \frac{g}{1\cdot 2}(\mathbf{r}\cdot\nabla)^2\phi(\mathbf{r}) + \cdots = \left[n + \frac{n(n-1)}{1\cdot 2}g + \cdots\right]\phi(\mathbf{r}).$$

This equation being true for an infinite number of values of g, we may equate the coefficients of the same powers of g on both sides of the equation, giving

$$\begin{aligned} n\phi(\mathbf{r}) &= \mathbf{r}\cdot\nabla\phi(\mathbf{r}), \\ n(n-1)\phi(\mathbf{r}) &= (\mathbf{r}\cdot\nabla)^2\phi(\mathbf{r}), \\ n(n-1)(n-2)\phi(\mathbf{r}) &= (\mathbf{r}\cdot\nabla)^3\phi(\mathbf{r}). \\ \cdot\quad\cdot\quad\cdot\quad&\cdot\quad\cdot\quad\cdot\quad\cdot \end{aligned} \tag{143}$$

The first of equations (143) is known as Euler's Theorem on homogeneous functions, which in terms of $x\ y\ z$ becomes

$$n\phi(x\,y\,z) = x\frac{\partial\phi}{\partial x} + y\frac{\partial\phi}{\partial y} + z\frac{\partial\phi}{\partial z}. \tag{144}$$

The remaining equations are extensions of the theorem, involving derivatives of higher orders.

Operators Involving ∇ Twice.

63. Possible Expressions Containing ∇ Twice. Given a scalar point-function V and a vector function $\mathbf{F}$, the following six combinations, involving ∇ twice, are possible ones:

1°	$\nabla\cdot\nabla V \equiv \nabla^2 V$	(div grad V, a scalar)
2°	$\nabla\times\nabla V$	(curl grad $V \equiv 0$)
3°	$(\nabla\cdot\nabla)\,\mathbf{F}$	($\nabla^2\mathbf{F}$, a vector).
4°	$\nabla(\nabla\cdot\mathbf{F})$	(grad div $\mathbf{F}$, a vector).
5°	$\nabla\cdot(\nabla\times\mathbf{F})$	(div curl $\mathbf{F} \equiv 0$).
6°	$\nabla\times(\nabla\times\mathbf{F})$	(curl curl $\mathbf{F} \equiv$ curl^2 $\mathbf{F}$, a vector.)

Two of these expressions vanish identically,

$$\nabla\times\nabla V \equiv 0, \qquad \text{curl (grad } V) \equiv 0$$

because any vector product containing two like vectors is zero, and

$$\nabla\cdot\nabla\times\mathbf{F} \equiv 0, \qquad \text{div (curl } \mathbf{F}) \equiv 0$$

because any triple scalar-product with two like vectors is also zero. These two results may be proved, if not sufficiently evident, by expanding according to the ordinary rules.

The 6th may be expanded into

$$\nabla\times(\nabla\times\mathbf{F}) = \nabla(\nabla\cdot\mathbf{F}) - \mathbf{F}(\nabla\cdot\nabla) \tag{145}$$

so that the 6th = 4th - 3d.

Equation (145) is sometimes written as

$$\text{curl (curl } \mathbf{F}) \equiv \text{curl}^2\,\mathbf{F} = \text{grad div } \mathbf{F} - \nabla^2\mathbf{F}. \tag{146}$$

This last important equation would be, as it is written, rather difficult to remember, but the advantage of retaining the notation in dels is made evident by the previous equation which may always be written out according to the rules for the expansion of a triple vector-product. As another example of the advantage of the symbolic notation it is perfectly easy to remember whether it is curl grad V or grad curl V, which is identically zero. For when they are written in terms of ∇, *i.e.*, $\nabla\times\nabla V$ and $\nabla(\nabla\times V)$, one is evidently zero and the other can have no meaning at all. **In other words, the del-notation when interpreted according to the ordinary rules of vector products, leads us to correct results independently of any physical or other considerations.**

The Operator ∇^2 or $\nabla\cdot\nabla$. (Read del square of . . .)

Operating on

$$\nabla V \equiv \mathbf{i}\frac{\partial V}{\partial x} + \mathbf{j}\frac{\partial V}{\partial y} + \mathbf{k}\frac{\partial V}{\partial z}$$

with $\nabla\cdot$, or in other words taking the div of the grad of V, we obtain,

$$\nabla\cdot\nabla V \equiv \nabla^2 V \equiv \frac{\partial^2 V}{\partial x^2} + \frac{\partial^2 V}{\partial y^2} + \frac{\partial^2 V}{\partial z^2} \equiv \left(\frac{\partial^2}{\partial x^2} + \frac{\partial^2}{\partial y^2} + \frac{\partial^2}{\partial z^2}\right)V, \quad (147)$$

the well known Laplacian operator, which when equated to zero is satisfied by the potential function in free space.

It is evident on inspection that

$$\nabla\cdot(\nabla V) = (\nabla\cdot\nabla) V \equiv \nabla^2 V.$$

Since the curl of grad V, or in symbols $\nabla\times\nabla V$, is zero identically there can be no ambiguity whatever when ∇ is twice applied to a scalar-function. When ∇^2 is applied to a vector function, it means that it is applied to the three scalar-function components of the vector function, and hence offers no new difficulty.

If

$$\nabla^2\mathbf{F} = 0$$

then

$$\nabla^2 F_1 = 0$$
$$\nabla^2 F_2 = 0$$
$$\nabla^2 F_3 = 0.$$

that is the three components of $\mathbf{F}$ satisfy Laplace's equation.

Since $$\nabla\times\nabla V = 0 = \text{curl grad } V,$$

it follows that the vector ∇V is a lamellar vector, by (§ 59).

Since $$\nabla\cdot\nabla\times\mathbf{F} = 0 = \text{div. curl } \mathbf{F},$$

it follows that the curl of any vector is a solenoidal vector, by (§ 53).

64. Differentiation of the Scalar Function r^m by ∇^2. We proved equation (108), that

$$\nabla r^m = mr^{m-1}\,\mathbf{r}_1 = mr^{m-2}\,\mathbf{r}.$$

Take the divergence ($\nabla\cdot$) of this vector.

$$\begin{aligned}\nabla\cdot\nabla\, r^m = \nabla^2 r^m &= m\,\{r^{m-2}\,\nabla\cdot\mathbf{r} + \mathbf{r}\cdot\nabla\, r^{m-2}\}\\ &= m\,\{3\,r^{m-2} + r(m-2)r^{m-3}\},\end{aligned}$$

because by (112),

$$\nabla\cdot\mathbf{r} = 3 \quad\text{and}\quad \mathbf{r}\cdot\nabla = r\;\;\mathbf{r}_1\cdot\nabla = r\frac{d}{dr}$$

so that $$\nabla^2 r^m = r^{m-2}\,\{3\,m + (m-2)\,m\},$$

and finally $$\nabla^2 r^m = m(m+1)r^{m-2}. \tag{148}$$

The two values of m which will satisfy the differential equation

$$\nabla^2 r^m = 0,$$

are easily seen to be $m = 0$ and $m = -1$, so that the scalar function $\frac{1}{r}$ satisfies Laplace's equation, or

$$\nabla^2\frac{1}{r} = 0.$$

This may also be shown, of course, by direct differentiation of the function $\frac{1}{r}$.

EXERCISES AND PROBLEMS.

1. Prove that $\nabla\cdot\mathbf{F}$ is an operator independent of choice of axes, by actually carrying out the transformation to a new set of axes. If the coördinates of the new set be denoted by primes, it should be found that

$$\frac{\partial X}{\partial x}+\frac{\partial Y}{\partial y}+\frac{\partial Z}{\partial z}=\frac{\partial X'}{\partial x'}+\frac{\partial Y'}{\partial y'}+\frac{\partial Z'}{\partial z'},$$

where X,Y,Z, are the components of a vector function **F**, and where X′,Y′,Z′ are its components referred to the new axes.

2. Prove directly by a change of axes that

$$\nabla\times\mathbf{F} \quad \text{or} \quad \text{Curl}\ \mathbf{F}$$

is invariant to that change.

3. Prove that

$$\nabla\mathbf{a}\cdot\mathbf{r}=\mathbf{a},$$
$$\nabla r=\mathbf{r}_1,$$
$$\nabla\mathbf{r}^2=\nabla\mathbf{r}\cdot\mathbf{r}=2\,\mathbf{r},$$

where a is a constant vector. These follow from the relation

$$d\,(\ )=d\mathbf{r}\cdot\nabla\,(\ ).$$

4. Prove that

$$\nabla\cdot\mathbf{r}=3,$$
$$\nabla\cdot\mathbf{r}_1=\frac{2}{r}, \qquad \nabla\cdot(\mathbf{a}\times\mathbf{r})=0.$$
$$\nabla\cdot\frac{1}{\mathbf{r}}=\frac{1}{r^2}.$$

5. Verify that

$$\nabla\cdot\nabla\mathbf{F}=\left(\frac{\partial^2}{\partial x^2}+\frac{\partial^2}{\partial y^2}+\frac{\partial^2}{\partial z^2}\right)\mathbf{F},$$

where $\mathbf{F}=\mathbf{i}X+\mathbf{j}Y+\mathbf{k}Z.$

6. Prove that

$$(\mathbf{a}\cdot\nabla)V=\mathbf{a}\cdot(\nabla V)$$

and state the resulting theorem. Apply this to a simple problem in potential.

7. Show that

$$\mathbf{a}\cdot\nabla\left(\frac{1}{r}\right) = -\frac{\mathbf{a}\cdot\mathbf{r}}{r^3} = -\frac{\mathbf{a}\cdot\mathbf{r}_1}{r^2},$$

$$\mathbf{b}\cdot\nabla\left(\mathbf{a}\cdot\nabla\frac{1}{r}\right) = \frac{3\,\mathbf{a}\cdot\mathbf{r}\;\mathbf{b}\cdot\mathbf{r}}{r^5} + \frac{\mathbf{a}\cdot\mathbf{b}}{r^3},$$

where **a** and **b** are constant vectors.

8. Show by direct expansion that

$$\nabla\times\nabla V \equiv 0$$

and

$$\nabla\cdot\nabla\times V \equiv 0.$$

9. Find the resultant attraction at the origin of the masses 12, 16, and 20 units respectively concentrated at the ends of the vectors $\mathbf{a} = 3\,\mathbf{i} + 4\,\mathbf{j}$, $\mathbf{b} = -5\,\mathbf{i} + 12\,\mathbf{j}$, and $\mathbf{c} = 8\,\mathbf{i} - 6\,\mathbf{j}$.

10. From the expression for the attraction at a point P due to a mass M, its density ρ at any point being a point-function of **r**,

$$\mathbf{F} = \iiint_{\text{vol}} \frac{\rho\,\mathbf{r}_1 dv}{r^2};$$

deduce the ordinary Cartesian expressions for the component attractions along the axes.

11. What theorems do equations (111), (117), (118) express? Write them out in Cartesian notation.

12. Explain paragraphs 46 and 47, considering instead of potential

(*a*) Temperature distribution in a body.
(*b*) Velocity distribution in a fluid.

CHAPTER VI.

APPLICATIONS TO ELECTRICAL THEORY.

65. Gauss's Theorem. Solid Angle. Consider a point 0, and any small area in space dS. Join every point in the boundary of the area dS to 0, thus forming a small cone. We define as the *solid angle* subtended by the area dS at 0, the value of $\frac{d\Sigma}{r^2}$, where $d\Sigma$ is the area cut out, by the cone, on any sphere of radius r, described about 0 as center. This is *numerically* equal to the area $d\omega$, cut out by the same cone on

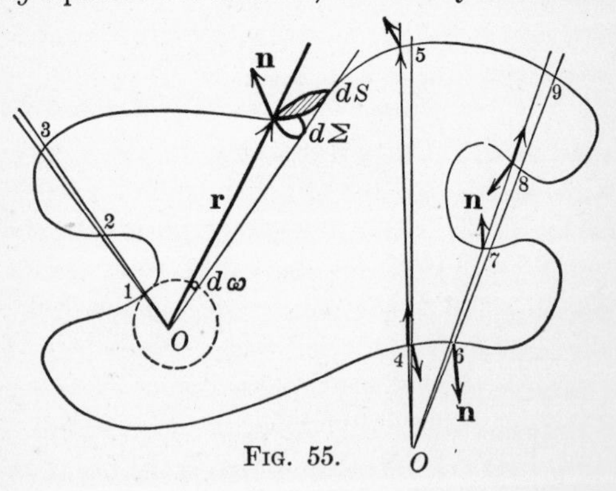

FIG. 55.

the sphere of unit radius, described about 0. The dimensions of solid angle are evidently zero. Since the total area of a unit sphere is equal to 4π, this is also the solid angle subtended by the whole of space or by *any surface* which completely surrounds the point O. $d\Sigma$ evidently subtends the same solid angle at O as dS. Calling $\mathbf{n}$ the unit normal to dS, its sense being taken in any conventional manner previously agreed upon (outward from the surface in the following),

$$d\Sigma = \pm\, dS \cos(\mathbf{rn}),$$

according as it makes an acute or an obtuse angle with **r**, respectively.

Now because $d\omega$ and $d\Sigma$ are parallel sections of the same cone we may write

$$d\Sigma = r^2 d\omega,$$

so that the solid angle

$$d\omega = \frac{d\Sigma}{r^2} = \pm \frac{dS \cos(\mathbf{rn})}{r^2}. \tag{149}$$

If a point O is chosen inside any closed surface, any small cone with vertex at O will cut out through the surface always once more than it cuts into it, but if the point be outside the surface it will cut in as many times as it cuts out. If **n** be chosen positive when drawn outwards, then the angle between **n** and **r** will be acute wherever the cone cuts out, obtuse wherever it cuts in. Thus an elementary cone when its vertex is inside a closed surface contributes an element of solid angle, $+d\omega$. As for example, in the figure the solid angles $+ d\omega$ and $- d\omega$, due to 1 and 2, annul each other, leaving $+ d\omega$ due to 3. When the vertex is outside, the resulting solid angle is zero; as for example, the solid angles $-d\omega$ and $+ d\omega$, due to 4 and 5, completely annul each other. So also do the elements at 6 and 7 and 8 and 9 in pairs. If we integrate or sum up all the solid angles due to all the elementary areas dS over the whole surface S, we shall obtain 4π for the sum, if the point O is inside and zero if the point is outside.

Expressing these results in symbols we have

$$\iint d\omega = \iint_S \frac{\cos(\mathbf{rn})}{r^2}\, dS = \iint_S \frac{\mathbf{n}\cdot\mathbf{r}}{r^3}\, dS = 4\pi \qquad O \text{ inside of } S \tag{150}$$

$$= \iint_S \frac{\cos(\mathbf{rn})}{r^2}\, dS = \iint_S \frac{\mathbf{n}\cdot\mathbf{r}}{r^3}\, dS = 0 \qquad O \text{ outside of } S.$$

These results, which are purely mathematical, are known as Gauss's Theorem.

Gauss's Theorem for the Plane. In a plane we may obtain an analogous theorem, *i.e.*, the plane angle subtended by a closed contour in a plane at a point O is 2π or O according as the point is inside or outside of the closed contour. In the figure (56) consider a point A connected to O by a radius vector which starts from B and moves once around the contour until it reaches B again. Evidently whatever the shape

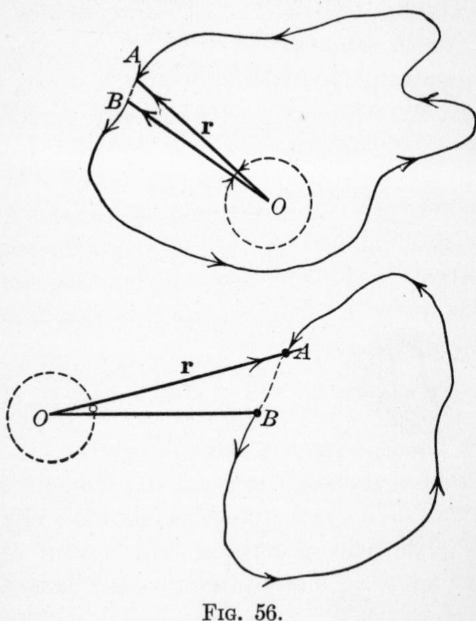

Fig. 56.

of the contour may be, the radius vector **r** has made but one revolution about O and therefore covered an arc equal to 2π on the unit circle about O. In the case that the contour is completely outside of O the same reasoning shows that the radius vector when it reaches B has not rotated *around* O at all.

In a plane the angle $d\psi$ subtended by an arc $d\mathbf{r}$ at a point O is, using a notation similar to (159),

$$d\psi = \pm \frac{dr \cos (\mathbf{rn})}{r} = \pm \frac{\mathbf{n}\cdot d\mathbf{r}}{r},$$

and the integrals are

$$\int_{\mathfrak{C}} d\psi = \int_{\mathfrak{C}} \frac{\cos(\mathbf{rn})}{r}\,dr = \int_{\mathfrak{C}} \frac{\mathbf{n}\cdot d\mathbf{r}}{r} = 2\pi \text{ or } = O$$

according as the point O is inside or outside of the contour.

We may apply analogous reasoning to Gauss's Theorem in space. Think of the solid angle subtended by an extensible sheet which gradually is extended completely in any manner around any point O, or which is made to form a closed surface of any shape completely outside of O, and the two results of equation (150) will become visually self-evident.

The importance of this theorem in physics is the fact that the surface-integral of the normal-component of the very important vector-function varying inversely as the square of the distance $\mathbf{r}$ from a point O is in symbols

$$\iint_S \frac{\cos(\mathbf{rn})}{r^2}\,dS = \iint \frac{\mathbf{n}\cdot\mathbf{r}_1}{r^2}\,dS,$$

and Gauss's Theorem reduces these integrals.

Second Proof of Gauss's Theorem. We may obtain another proof of Gauss's Theorem from the following physical considerations. The field of force around a point at which is concentrated a unit of matter, electrical say, is by Coulomb's Law

$$\mathbf{F} \propto \frac{\mathbf{r}_1}{r^2}. \tag{151}$$

This means that the vector $\mathbf{F}$ is directed radially outwards, but that its magnitude varies inversely as r^2. Consider two closed surfaces S_1 and S_2, the first surrounding the point O and the other lying completely outside of O. About O draw two spheres, one of which is completely outside of both of the two surfaces S_1 and S_2, and the other of a small enough radius so that it does not touch either S_1 or S_2. Since the magnitude of **F** falls off exactly as fast as the areas of the spheres increase with increasing radius, the surface integral of this vector over each of the spheres is the same. Or, in other

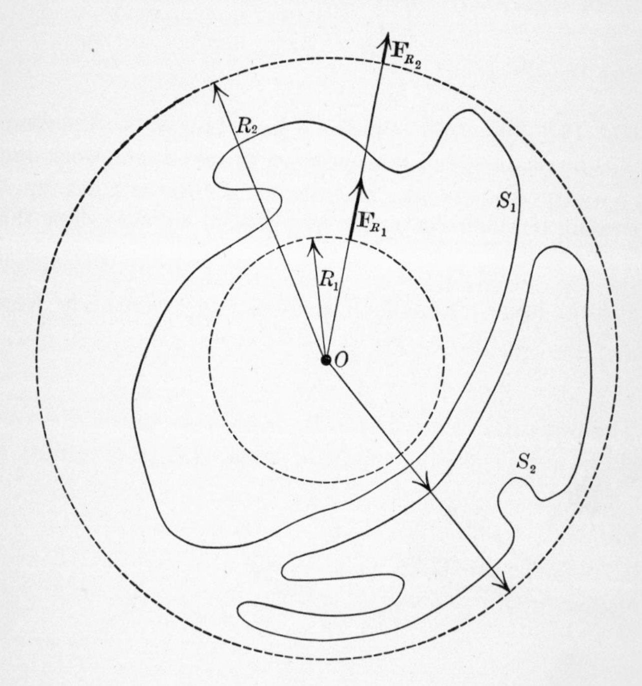

Fig. 57.

words, the flux that gets through one reaches the other. Evidently the same amount of flux must have passed through the surface S_1, so that we may write for the flux

$$\int\int_{S_1} \frac{\mathbf{n} \cdot \mathbf{r}_1}{r^2} dS = \frac{1}{R_1^2} \int\int_{\substack{\text{sphere} \\ \text{of rad. } R_1}} \mathbf{r}_1 \cdot \mathbf{r}_1 \, dS = \frac{1}{R_1^2} \int\int dS$$

$$= \frac{1}{R_1^2} 4 \pi R_1^2 = 4 \pi.$$

As for the surface S_2, since no flux is gained or lost in going from sphere R_1 to sphere R_2, whatever flux went into S_2 must also have come out of it again, and therefore

$$\int\int_{S_2} \frac{\mathbf{n} \cdot \mathbf{r}_1}{r^2} dS = 0.$$

Hence the theorem.*

66. The Potential. Poisson's and Laplace's Equations. From the definition of the potential in § 46 as the work done on a unit positive charge in bringing it from infinity up to the point at which the potential is desired, we may show that the scalar function $\frac{m}{r}$ is the potential function corresponding to the inverse square, or Coulomb's (in gravitation, Newton's) law of force $F \propto \frac{m\,\mathbf{r}_1}{r^2}$.

Consider a quantity of positive matter, m, at O.

The potential at any point P, or in other words the work done *on* a unit positive charge in bringing it from infinity to

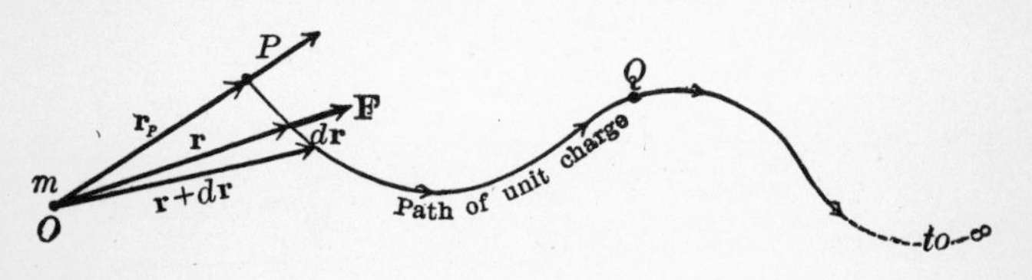

FIG. 58.

P, is equal to the line-integral of the force function F from P to ∞ along the path PQ traversed by the unit charge.

$$V_P = \int^{\infty} \mathbf{F} \cdot d\mathbf{r} = -\int_{\infty}^{r_P} \frac{m}{r^2} \mathbf{r}_1 \cdot d\mathbf{r}, \qquad \text{see §16}$$

which may be written, because $\nabla \frac{1}{r} = -\frac{\mathbf{r}_1}{r^2}$ by (109),

$$V_P = \int_{\infty}^{r_P} d\mathbf{r} \cdot \nabla \left(\frac{m}{r}\right). \qquad \textbf{(152)}$$

* See Appendix, p. 251, for another proof of Gauss's Theorem.

This last expression is the total derivative of $\frac{m}{r}$, so that the integral is independent of the path and depends only upon the limits, that is, upon the starting point and ending point, giving

$$V_P = \frac{m}{r}\Big/_{\infty}^{r_P} = \frac{m}{r_P}.$$

Should there be other masses present, the potential function

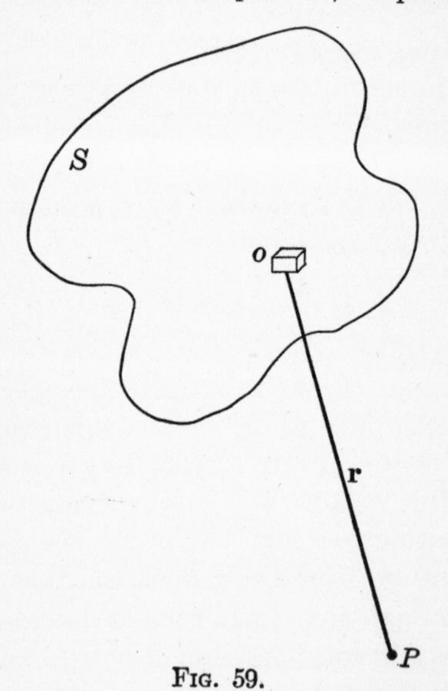

FIG. 59.

due to them all is the sum of the separate functions due to each, or

$$V = \frac{m_1}{r_{P_1}} + \frac{m_2}{r_{P_2}} + \cdots = \sum_1^s \frac{m_s}{r_{P_s}}, \tag{153}$$

where $\mathbf{r}_{P_s}$ is the distance from the point at which the potential is to be found to the mass m_s. If the masses instead of being at discrete points form a continuous distribution, the summation becomes a volume integral; *dm*, the element of mass, becomes $\rho\, dv$, where ρ is the volume density of the matter under consideration. We have then,

$$V = \iiint_{\text{mass}} \frac{dm}{r} = \iiint_{\text{vol}} \frac{\rho\, dv}{r}. \tag{154}$$

The integral thus defined which is to be taken over the volume occupied by the masses may be shown to be finite, continuous, uniform, as well as its first derivatives. It vanishes itself at infinity to the first order, and its first derivatives to the second order.

A system of forces for which the line-integral between any two points is independent of the path is called a **Conservative System.**

If we multiply Gauss's Integrals by m, a mass concentrated at the point O, we shall obtain

$$\iint_S \frac{m}{r^2}\, \mathbf{n} \cdot \mathbf{r}_1 dS = 4\,\pi m. \tag{155}$$

This integral states that the outward normal component of flux of force (according to the inverse square or Newtonian law) through a closed surface surrounding O is $4\,\pi$ times the amount of matter within; any matter lying outside of the surface contributing nothing. As every element of mass m contributes $4\,\pi m$ we have the proposition that the outward flux of force through any closed surface due to any distribution is 4π times the total amount of matter within the surface. It is in this form that Gauss's Integral is usually given, but evidently, from what precedes, it is a geometrical theorem rather than an electrical or gravitational one.

The force at any point due to any distribution of matter is $-$ grad V or $- \nabla V$, by § 47, where

$$V = \iiint_\infty \frac{\rho\, dv}{r}$$

is the potential function due to the distribution. The sign ∞ denotes that the integral is to be taken over the whole of space. This is equivalent to integrating over the matter *alone*, as wherever there is no matter $\rho = 0$ and the integral contributes nothing. So that we may write (155) as

$$-\iint_S \mathbf{n}\cdot\nabla V\, dS = 4\,\pi \iiint_S \rho\, dv,$$

$\iiint_S \rho\, dv$ being the total quantity of matter within S. By means of the divergence theorem (121) the surface integral above may be transformed into a volume integral taken throughout the volume enclosed by S.

$$-\iint_S \mathbf{n}\cdot\nabla V\, dS = -\iiint_S \nabla\cdot\nabla V\, dv = 4\,\pi \iiint_S \rho\, dv.$$

As this equality holds whatever surface S is taken, it follows that the integrands are everywhere equal and

$$\nabla\cdot\nabla V = \nabla^2 V = -\,4\,\pi\rho. \tag{156}$$

This is **Poisson's Equation.**

In free space where $\rho = 0$ this becomes

$$\nabla^2 V = 0, \tag{157}$$

which is **Laplace's Equation.**

We may interpret these equations as follows: Every quantity of matter emits lines of force, $4\,\pi$ lines per unit quantity. This numeric $4\,\pi$ is purely conventional and appears because the intensity at unit distance from unit charge is defined as unity; and since unit intensity corresponds to one line per unit area, there must be $4\,\pi$ lines emitted in order to have one for each of the $4\,\pi$ units of area in the surface of the unit sphere. So then, if a surface of volume dv be drawn around a point where the density is ρ, the lines passing through the surface are equal to $4\,\pi$ times the quantity of matter $\rho\, dv$ within it, or

$$\text{div}\,\mathbf{F}\ (\text{due to volume } dv) = 4\,\pi\rho\, dv,$$

and per unit volume

$$\text{div}\,\mathbf{F} = 4\,\pi\rho.$$

Now if **F** have a potential, that is, if **F** can be represented as the grad or ∇ of some scalar function W, then putting $\mathbf{F} = \nabla W$,

$$\nabla \cdot \nabla W = \nabla^2 W = 4\pi\rho.$$

This equation is true for the potential W due to attracting matter.

In the case of repelling forces, since the force is opposite to the direction of increase in the scalar function, we may place $W = - V$, and we may write as before

$$\nabla^2 V = - 4\pi\rho.$$

In free space where there is no density ρ the equation becomes

$$\text{div}\, \mathbf{F} = 0,$$

which says that the lines of force are solenoidally distributed, that is, the flux takes place in unbroken continuous paths, and hence cannot end nor begin at any point of space devoid of matter.

The reason for the term (div) divergence is evident from the foregoing.

Harmonic Function. A function which in a region is single-valued, continuous, and satisfies Laplace's equation is said to be *harmonic* in that region.

A Spherical Harmonic of degree n is any *homogeneous* (143) harmonic point-function of space. That is, if V satisfies the equations

$$\nabla^2 V = 0$$

and

$$\mathbf{r} \cdot \nabla V = nV,$$

it is a spherical harmonic of degree n. The study and use of such functions is of great importance in all branches of mathematical physics.

67. Green's Theorems. Two theorems due to Green of very important application in theoretical physics follow immediately by an application of the divergence theorem,

$$\iint_S \mathbf{n} \cdot \mathbf{W} \, dS = \iiint_{\text{vol}} \nabla \cdot \mathbf{W} \, dv,$$

to the function $\mathbf{W} = U \nabla V$, where U and V are two scalar point-functions which with their derivatives are uniform and continuous in the space considered. Applying $\nabla \cdot$ to $\mathbf{W}$,

$$\nabla \cdot \mathbf{W} = \nabla \cdot (U \nabla V) = U \nabla^2 V + \nabla U \cdot \nabla V.$$

Substituting in the equation above

$$\iint \mathbf{n} \cdot U \nabla V \, dS = \iiint U \nabla^2 V \, dv + \iiint \nabla U \cdot \nabla V \, dv. \tag{158}$$

Similarly, by symmetry, putting for $\mathbf{W} = V \nabla U$, we have

$$\iint \mathbf{n} \cdot V \nabla U \, dS = \iiint V \nabla^2 U \, dv + \iiint \nabla U \cdot \nabla V \, dv.$$

Subtracting these two equations there remains

$$\iint \mathbf{n} \cdot (U \nabla V - V \nabla U) \, dS = \iiint (U \nabla^2 V - V \nabla^2 U) \, dv. \tag{159}$$

The surface integrals are to be taken over the surfaces bounding the region under consideration, and the volume integrals throughout the volumes enclosed by these surfaces. Equations (158) and (159) are called Green's Theorem in its first and second forms respectively.

68. Green's Formulæ. Apply Green's Theorem in its second form to two functions U and V. Let U be the function $U = \frac{1}{r}$, and let V be the potential due to any distribution of matter. The region to be considered is the space lying between the infinite sphere, S_∞, any surfaces S which surround the distribution, and the infinitesimal sphere of radius ε, surrounding O, the point from which $\mathbf{r}$ is measured.

The equation

$$\iint \mathbf{n}\cdot(U\nabla V - V\nabla U)\,dS = \iiint (U\nabla^2 V - V\nabla^2 U)\,dv \tag{160}$$

becomes, since $\nabla^2 U = \nabla^2 \frac{1}{r} = 0,$ by (148)

$$\iint \mathbf{n}\cdot\left(\frac{1}{r}\nabla V - V\nabla\frac{1}{r}\right)dS = \iiint \left(\frac{1}{r}\nabla^2 V\right)dv.$$

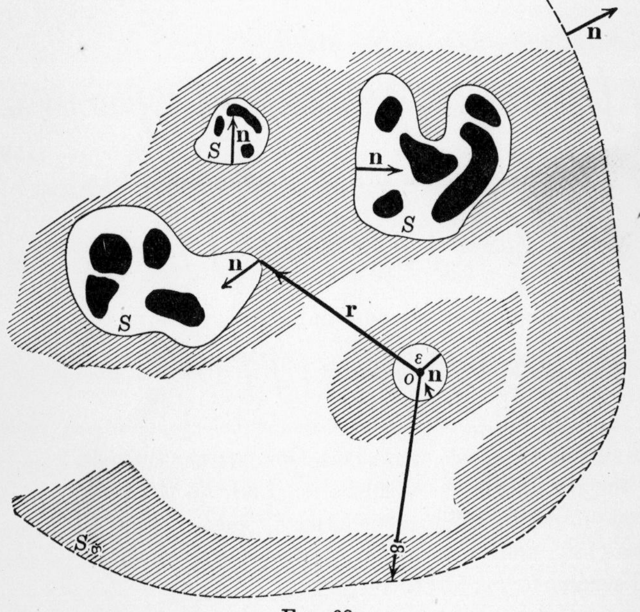

FIG. 60.

The surface integral is to be taken over the bounding surfaces S to the region. The infinite sphere contributes nothing, as at infinity $\frac{1}{r}\nabla V$ and $V\nabla\frac{1}{r}$ become zero to the third order, both containing r^3 in the denominator. For the small sphere about O the first part of the integral may be transformed,

$$\iint \mathbf{n} \cdot \frac{1}{r} \nabla V \, dS = \iint \frac{1}{\varepsilon} (\mathbf{n} \cdot \nabla V) \, \varepsilon^2 d\omega = \varepsilon \iint \mathbf{n} \cdot \nabla V \, d\omega,$$

where $d\omega$ is the solid angle subtended by an element of the small sphere at O, and, where ε, its small radius, is constant during integration. As ε becomes smaller and smaller and because $\mathbf{n} \cdot \nabla V$, the normal force on the surface, is finite, the integral vanishes in the limit.

Considering the second part of the surface integral over the small sphere, we may write

$$- \iint \left(\mathbf{n} \cdot V \nabla \frac{1}{r} \right) dS = - V_0 \iint d\omega = - V_0 \times 4 \pi,$$

because $+ \mathbf{n} \cdot \nabla \frac{1}{r} dS = \frac{\mathbf{n} \cdot \mathbf{r}_1}{r^2} dS$ = solid angle due to dS. As the radius of the small sphere diminishes V approaches V_0, its value at O.

So that finally,

$$V_0 = - \frac{1}{4 \pi} \underset{\text{Region}}{\iiint} \frac{\nabla^2 V}{r} dv + \frac{1}{4 \pi} \underset{\text{Surfaces}}{\iint} \mathbf{n} \cdot \left(\frac{1}{r} \nabla V - V \nabla \frac{1}{r} \right) dS, \qquad (161)$$

the surface integral being taken over the bounding surfaces S and the volume integral over the region bounded by them, shaded in the figure.

If $\nabla^2 V$ is equal to zero in the region considered, the potential function at any point O is

$$V_0 = \frac{1}{4 \pi} \iint_S \mathbf{n} \cdot \left(\frac{1}{r} \nabla V - V \nabla \frac{1}{r} \right) dS, \qquad (162)$$

which shows that it is completely determined everywhere if the values of the potential V and of its normal derivative $\mathbf{n} \cdot \nabla V$ are known over the bounding surfaces S. If the matter producing this potential and distributed in any manner within S be taken out and replaced by a surface density of matter σ on S of amount

$$\sigma = \frac{1}{4\pi}\mathbf{n}\cdot\left(\nabla V - rV\nabla\frac{1}{r}\right), \tag{163}$$

the potential at O will be exactly the same as before, because by substituting this value for σ in

$$V_0 = \iint_S \frac{\sigma\, dS}{r}$$

we obtain equation (162). We shall call this distribution an **Equivalent Layer.** But in general this will not necessarily make the surface an equipotential surface.

If the point O is inside the surface S, a similar deduction gives the formula

$$V_0 = -\frac{1}{4\pi}\iiint \frac{\nabla^2 V}{r} dv + \frac{1}{4\pi}\iint \mathbf{n}\cdot\left(\frac{1}{r}\nabla V - V\nabla\frac{1}{r}\right) dS, \tag{164}$$

where $\mathbf{n}$ is to be drawn as the *external* normal to the region in which O lies. With this convention the two formulæ due to Green (161) and (164) are identical in form.

Green's Function. Adding together Green's equation (160), which may be written

$$0 = \iiint (U\nabla^2 V - V\nabla^2 U)\, dv - \iint \mathbf{n}\cdot(U\nabla V - V\nabla U)\, dS,$$

and (161), which hold under the same conditions, we obtain

$$4\pi V_0 = \iiint \left(U - \frac{1}{r}\right)\nabla^2 V\, dv - \iiint V\nabla^2 U\, dv$$
$$- \iint \left[\left(U - \frac{1}{r}\right)\nabla V - V\nabla\left(U - \frac{1}{r}\right)\right]\cdot\mathbf{n}\, dS. \tag{165}$$

This equation is of especial importance in the theories of light and electricity. The quantity $\left(U - \frac{1}{r}\right)$ which appears in the integral is sometimes known as Green's Function.

69. Solution of Poisson's Equation. Equation (161) states that if the quantity $\nabla^2 V$ is known throughout a region bounded by any surface S, and if the quantities V and ∇V are known at all points of the surface, then V is completely determined within the surface. Allow the surface S to recede to infinity so that we are now considering the potential in the whole of space. Then in the equation the surface integral contributes nothing, as all the quantities multiplied by dS approach zero to a sufficiently high order. There remains, then, in the limit only

$$V_0 = -\frac{1}{4\pi} \int\int\int_\infty \frac{\nabla^2 V}{r}\, dv. \tag{166}$$

Now, by Poisson's Equation, V satisfies the relation

$$\nabla^2 V = -4\pi\rho.$$

So that, by (166), if the value of ρ, the density, be given at every point in space, V is determined by the integral

$$V = \int\int\int_\infty \frac{\rho\, dv}{r}, \tag{167}$$

which is therefore a solution of Poisson's Equation.

The Integrating Operator Pot. The operation of finding the potential due to a distribution whose density is defined everywhere by the scalar function $\rho(\mathbf{r})$ plays such an important rôle in mathematical physics that Prof. Gibbs has given to this operation a special name and defines

$$\text{Pot}\, \rho \equiv \int\int\int_\infty \frac{\rho\, dv}{r} \tag{168}$$

(read potential of ρ).

The sign ∞ indicates that the limits may be taken over the whole of space, as wherever there is no matter the integral contributes nothing. We shall call the operation indicated by the above equation "the potential of ρ," even when ρ does not represent a volume density.

Considering again equation (166)

$$V = -\frac{1}{4\pi}\iiint \frac{\nabla^2 V}{r}\,dv,$$

and using the notation (168), we see that we may write

$$V = -\frac{1}{4\pi}\operatorname{pot}\nabla^2 V. \qquad (169)$$

So that the application of

$$-\frac{1}{4\pi}\operatorname{pot}(\quad),$$

to a function nullifies the effect of ∇^2 on that same function, or, in other words,

$$-\frac{1}{4\pi}\operatorname{pot}(\quad) \text{ is the inverse operator to } \nabla^2(\quad).$$

70. Vector-Potential. In the same way that the potential due to the scalar function ρ is formed, that is,

$$V = \operatorname{pot}\rho = \iiint_\infty \frac{\rho}{r}\,dv,$$

we may *define* the potential of the vector function

$$\boldsymbol{\rho} = \rho_1\mathbf{i} + \rho_2\mathbf{j} + \rho_3\mathbf{k},$$

where ρ_1, ρ_2 and ρ_3, are given scalar functions of $\mathbf{r}$, as

$$\mathbf{V} = \operatorname{pot}\boldsymbol{\rho} = \iiint_\infty \frac{\boldsymbol{\rho}}{r}\,dv$$

$$= \mathbf{i}\iiint_\infty \frac{\rho_1}{r}\,dv + \mathbf{j}\iiint_\infty \frac{\rho_2}{r}\,dv + \mathbf{k}\iiint_\infty \frac{\rho_3}{r}\,dv.$$

(read vector-potential of $\boldsymbol{\rho}$)

The vector function **V** so defined is called the *vector-potential* of **ρ**. Its three components evidently satisfy the relations satisfied by the scalar potential, so that we have

$$\nabla^2 \mathbf{V} = -4\pi \boldsymbol{\rho}. \tag{170}$$

In strict analogy with the solution of Poisson's Equation for a scalar potential we have then for the solution of a vector-potential

$$\mathbf{V} = -\frac{1}{4\pi}\iiint_\infty \frac{\nabla^2 \mathbf{V}}{r}\, dv. \tag{171}$$

71. Separation of a Vector Point-Function *W*, which has a Vector-Potential, into Solenoidal or Rotational and Lamellar or Irrotational Components. This means that the vector function **W** is to be separated into two parts, one of which has no divergence and the other no curl. We then assume

$$\mathbf{W} = \mathbf{X} + \mathbf{Y}, \tag{172}$$

where $\nabla\cdot\mathbf{X} = 0$ and $\nabla\times\mathbf{Y} = 0.$

Consider the scalar function ϕ and the vector function **V**, related to **X** and **Y**, respectively, in the following manner:

$$\mathbf{X} = \nabla\times\mathbf{V},$$

$$\mathbf{Y} = -\nabla\phi.$$

Then

$$\mathbf{W} = \nabla\times\mathbf{V} - \nabla\phi.$$

If it is possible to determine **V** and ϕ the problem is solved. To do this take the divergence of **W**, giving

$$\nabla\cdot\mathbf{W} = -\nabla^2\phi.$$

So that the solution for ϕ is, by (166),

$$\phi = -\frac{1}{4\pi}\iiint\frac{\nabla^2\phi}{r}\, dv = \frac{1}{4\pi}\iiint\frac{\nabla\cdot\mathbf{W}}{r}\, dv \tag{173}$$

$$= \frac{1}{4\pi}\,\text{pot}\,(\nabla\cdot\mathbf{W}).$$

Similarly, taking the curl of **W**,

$$\nabla\times\mathbf{W} = \nabla\times(\nabla\times\mathbf{V}) = \nabla(\nabla\cdot\mathbf{V}) - \nabla^2\mathbf{V}.$$

Now since **V** is as yet undetermined, we may assume that its divergence is zero or that $\nabla\cdot\mathbf{V} = 0$, hence

$$\nabla\times\mathbf{W} = -\nabla^2\mathbf{V}.$$

So that $\nabla\times\mathbf{W}$ is 4π times the vector function of which **V** is the vector potential, and by (171)

$$\mathbf{V} = -\frac{1}{4\pi}\iiint\frac{\nabla^2\mathbf{V}}{r}\,dv = \frac{1}{4\pi}\iiint\frac{\nabla\times\mathbf{W}}{r}\,dv$$

$$= \frac{1}{4\pi}\text{pot}\,(\nabla\times\mathbf{W}), \text{ which determines } \mathbf{V}. \tag{174}$$

Finally, since

$$\mathbf{W} = \nabla\times\mathbf{V} - \nabla\phi,$$

$$\mathbf{W} = -\frac{1}{4\pi}\nabla\iiint\frac{\nabla\cdot\mathbf{W}}{r}\,dv + \frac{1}{4\pi}\nabla\times\iiint\frac{\nabla\times\mathbf{W}}{r}\,dv, \tag{175}$$

the decomposition is thus accomplished. This decomposition is sometimes known as **Helmholtz's Theorem.***

Other Systems of Units. The factor 4π which occurs in many of these equations is due to the definition of unit quantity of matter. In virtue of this definition it is necessary to assume that every unit of matter emits 4π lines of force, so that, for example, the number of lines cutting through any closed surface around any amount of matter will be 4π times as many as there are units of matter inside. Of late it is the fashion to eliminate this "eruption of π's" as **Heaviside** has it. This may be done in various ways, one of which is to redefine the unit quantity in such a manner that it emits but one line of force, in which case the equations

$$\mathbf{F} = \frac{m_1 m_2}{r^2}\mathbf{r}_1 \quad \text{div}\,\mathbf{F} = 4\pi\rho \quad \text{and} \quad \nabla^2 V = -4\pi\rho,$$

become respectively

$$\mathbf{F} = \frac{1}{4\pi}\frac{m_1 m_2}{r}\mathbf{r}_1 \quad \text{div}\,\mathbf{F} = \rho \quad \text{and} \quad \nabla^2 V = -\rho. \tag{176}$$

* Wiss. Abh. Band I, p. 101. For a mnemonic if not another proof of this theorem, multiply equation (145) by $\frac{1}{4\pi}\frac{1}{r}$ and integrate, remembering equation (171).

Such a choice of units eliminates 4π in a number of formulæ but introduces it in others. It is nevertheless the most convenient assumption in the modern theory, where the energy is located in the space between the acting matter, and in which action at a distance no longer holds first place. The potential at a distance r from a mass m, for example, becomes $\frac{1}{4\pi}\frac{m}{r}$ instead of $\frac{m}{r}$. The operation of forming the potential, or pot V, would in this system consist in forming the integral,

$$\text{pot}\,\rho = \iiint \frac{\rho\,dv}{4\,\pi r}, \tag{177}$$

and the theorem of Helmholtz would become in this notation

$$\mathbf{W} = -\nabla\,\text{pot}\,(\nabla\cdot\mathbf{W}) + \nabla\times\text{pot}\,(\nabla\times\mathbf{W}), \tag{178}$$

72. Energy of a System in Terms of Potential. Consider two particles of matter acting according to the inverse

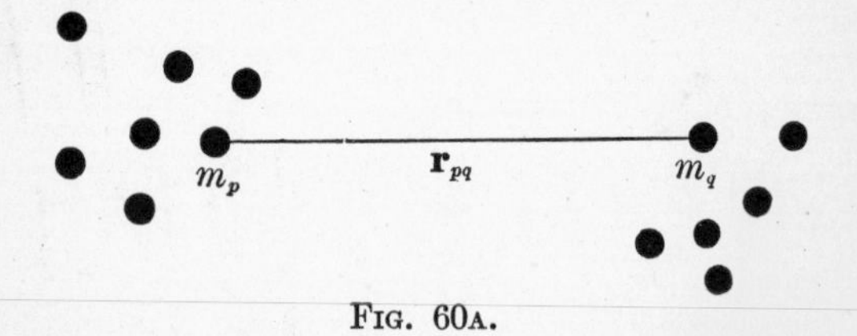

FIG. 60A.

square law, m_p and m_q, respectively, separated by a distance r_{pq}. In order to bring the mass m_p from infinity to its position an amount of work (§ 66)

$$W_{pq} = \frac{m_q}{r_{pq}} m_p,$$

must be expended *on* the mass m_p if the masses repel, or *by* m_p if they attract. For definiteness assume the matter to be repelling.

The expression above may be written in two ways,

$$W_{pq} = V_q m_p \quad \text{or} \quad V_p m_q$$

because
$$V_q = \frac{m_q}{r_{pq}} \quad \text{and} \quad V_p = \frac{m_p}{r_{pq}}.$$

Similarly if we have any two systems of particles, the energy obtainable by allowing the two systems to disperse to an infinite distance apart is

$$W_{pq} = \Sigma_p \Sigma_q \frac{m_p m_q}{r_{pq}},$$

where the summation signs extend to every pair of points, one point from each of the systems. If we consider the two systems as one, a factor $\frac{1}{2}$ must be introduced, as in the summation every term would appear twice, so that

$$W_{pq} = \frac{1}{2} \Sigma_p \Sigma_q \frac{m_p m_q}{r_{pq}} \tag{179}$$

represents the mutual potential energy of a *single* system of particles. If the system forms a continuous distribution the summation (179) becomes

$$W = \frac{1}{2} \int\int\int_\infty \int\int\int_\infty \frac{dm\, dm'}{r} = \frac{1}{2} \int\int\int_\infty V\, dm,$$

where V is the potential function due to the total distribution.

73. Energy of a Distribution in Terms of Field Intensity. If the distribution consists of a surface and a volume distribution of surface density σ and volume density ρ, the above integral takes the form

$$W = \frac{1}{2} \int\int_S V \sigma\, dS + \frac{1}{2} \int\int\int_\infty V \rho\, dv. \tag{180}$$

The integrals are taken over the surfaces and throughout the volumes of the matter under consideration respectively.

At a surface distribution there is a discontinuity or change in the normal component of the force due to the *surface* distribution, given by the well-known expression

$$F_n = 4\,\pi\sigma = \mathbf{n}\cdot\nabla V.$$

By drawing a surface completely surrounding the surface distributions we may apply Green's Theorem to the whole of space outside of these surfaces. Remembering that from the last equation

$$\sigma = \frac{1}{4\,\pi}\,\mathbf{n}\cdot\nabla V \quad \text{and that} \quad \rho = -\frac{1}{4\,\pi}\,\nabla^2 V,$$

the two integrals which are now

$$W = \frac{1}{8\,\pi}\int\!\!\int_S V\,\mathbf{n}\cdot\nabla V\,dS - \frac{1}{8\,\pi}\int\!\!\int\!\!\int_\infty V\,\nabla^2 V\,dv$$

become by Green's Theorem (158)

$$W = \frac{1}{8\,\pi}\int\!\!\int\!\!\int_\infty (\nabla V)^2\,dv = \frac{1}{8\,\pi}\int\!\!\int\!\!\int_\infty \mathbf{F}^2\,dv. \qquad (181)$$

If the medium is any other than vacuo, the element in the integral, $\mathbf{F}^2\,dv$ must be multiplied by a factor ε characteristic of the medium. The energy of the distribution is in this case

$$W = \frac{1}{8\,\pi}\int\!\!\int\!\!\int_\infty \varepsilon\,\mathbf{F}^2\,dv. \qquad (182)$$

This may also be written as

$$W = \frac{1}{8\,\pi}\int\!\!\int\!\!\int_\infty \varepsilon\,\mathbf{F}\cdot\mathbf{F}\,dv = \frac{1}{8\,\pi}\int\!\!\int\!\!\int \mathfrak{F}\cdot\mathbf{F}\,dv, \qquad (183)$$

where

$$\varepsilon\,\mathbf{F} \equiv \mathfrak{F} \qquad (184)$$

is a vector called the Induction.

74. Expressions for Surface and Volume Densities of a Distribution in Terms of the Intensity of Polarization. Starting again (180) with the energy of a surface and volume distribution of densities σ and ρ, respectively,

$$W = \frac{1}{2}\iint_S V\sigma\, dS + \frac{1}{2}\iiint_{\text{vol}} V\rho\, dv.$$

Let us assume that this energy may be written also as

$$W = -\frac{1}{2}\iiint \mathbf{I}\cdot\mathbf{H}\, dv = \frac{1}{2}\iiint \mathbf{I}\cdot\nabla V\, dv,$$

where **I** is called the intensity of polarization, **H** is the field strength, and V is the potential corresponding to **H**. Then since

$$\nabla\cdot(V\mathbf{I}) = \mathbf{I}\cdot\nabla V + V\nabla\cdot\mathbf{I},$$

the integral may be transformed into

$$W = \frac{1}{2}\iiint \nabla\cdot(V\mathbf{I})\, dv - \frac{1}{2}\iiint V\nabla\cdot\mathbf{I}\, dv.$$

Transforming the first integral by the divergence theorem and comparing this with the expression for the energy in terms of σ and ρ,

$$W = \frac{1}{2}\iint V(\mathbf{n}\cdot\mathbf{I})\, dS - \frac{1}{2}\iiint V\nabla\cdot\mathbf{I}\, dv,$$

we see that the polarization **I** produces a surface density

$$\sigma = \mathbf{n}\cdot\mathbf{I} \tag{185}$$

and a volume density

$$\rho = -\nabla\cdot\mathbf{I}.$$

Conversely, assuming a distribution to consist of a surface density σ and a volume density ρ, it is easy to show by reasoning backwards that there is a quantity **I** related to σ and ρ by the equations

$$\sigma = \mathbf{n}\cdot\mathbf{I} \quad \text{and} \quad \rho = -\nabla\cdot\mathbf{I}$$

such that the energy of the distribution may be represented by the integral throughout the volume,

$$W = -\frac{1}{2}\iiint \mathbf{I}\cdot\mathbf{H}\, dv. \tag{186}$$

Equations of the Electro-Magnetic Field.

75. Maxwell's Equations. By experiment **Faraday** showed that when the magnetic flux through a linear circuit is varied there is induced in the circuit an electro-motive force. If the circuit is a closed one this induced electro-motive force produces a current in it. He also showed that this electro-motive force is equal to the *negative* rate of change of the magnetic flux. The *positive* direction of rotation in a circuit is connected with the positive direction of flux through it, according to the adjoining diagram which symbolizes the so-called **cork-screw rule.** If the arrow shows direction of increase of magnetic flux, the arrow-head in circuit shows the direction opposite to the induced current. The figure as drawn shows the direction of the magnetic flux *due* to the current in the circuit. By Lenz's law such a magnetic flux would induce a current *opposite* to this; hence the negative sign in equation 187 below. Since electricity tends to flow from places of high to places of low potential we may consider this electro-motive force as something in the nature of an electrostatic field which is induced in the space by the varying flux. That is, the electro-motive force is induced in the space even when unoccupied by a conductor.

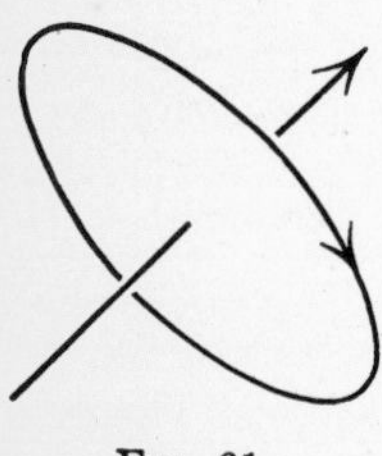

FIG. 61.

In a conductor this electro-motive force produces a current and in a non-conductor *tends* to produce a current. To obtain the total electro-motive force around any circuit, we evaluate the line-integral of this electrostatic field **F** along that circuit. Then the induced e.m.f. may be written

$$\int_{\circlearrowleft} \mathbf{F}\cdot d\mathbf{r} = \iint_{\text{cap}} (\nabla\times\mathbf{F})\cdot\mathbf{n}\, dS,$$

by Stokes' Theorem. But Faraday's experiment shows that

this is equal to the *negative* of the rate change of the magnetic induction $\mathcal{H}$ through this circuit, so that

$$\iint_{cap} (\nabla \times \mathbf{F}) \cdot \mathbf{n} \, dS = -\frac{d}{dt} \iint_{cap} \mathcal{H} \cdot \mathbf{n} \, dS = \iint_{cap} -\frac{d\mathcal{H}}{dt} \cdot \mathbf{n} \, dS.$$

As this is true whatever circuit is considered and whatever cap is taken as long as it is bounded by the circuit we may write

$$\nabla \times \mathbf{F} = -\frac{\partial \mathcal{H}}{\partial t}. \qquad (187)$$

By experiment **Ampère** proved that a current I is equivalent to any magnetic shell of a certain strength which is bounded by the current. He also showed that a current may be measured by means of the magnetic field that it produces, and quantitatively that 4π times the current in any section of a conductor is equal to the line integral of the magnetic force $\mathbf{H}$ taken *once* around any path linked positively with the conductor. If $\mathbf{q}$ be the current density, then, symbolically,

$$\int_{\mathfrak{H}} \mathbf{H} \cdot d\mathbf{r} = 4\pi \iint_{cap} \mathbf{q} \cdot \mathbf{n} \, dS,$$

where the surface integral is taken over any cap to the surface. Transforming the first integral by Stokes' Theorem, we have

$$\int_{\mathfrak{H}} \mathbf{H} \cdot d\mathbf{r} = \iint_{cap} \nabla \times \mathbf{H} \cdot \mathbf{n} \, dS = 4\pi \iint_{cap} \mathbf{q} \cdot \mathbf{n} \, dS,$$

and as this equation is true whatever portion of space is considered and whatever path is taken around that portion of space, we may write

$$\nabla \times \mathbf{H} = 4\pi \mathbf{q}. \qquad (188)$$

In order to explain the effect of an electro-motive force upon dielectric non-conductors **Maxwell** *assumed* that instead of a current $\mathbf{q}$ there is produced a so-called displacement-motion or current $\mathbf{q}'$ of electricity, which on the release of the inducing electro-motive force springs back and takes up its original position. He assumes that this current-displacement produces the same magnetic effect as would be produced by a current of density

$$\mathbf{q}' = \frac{1}{4\pi}\frac{\partial \mathfrak{F}}{\partial t} \tag{189}$$

where

$$\mathfrak{F} = \varepsilon \mathbf{F}$$

and ε is a constant of the medium called electric inductivity at every point of the field. We therefore, in considering a *dielectric*, introduce this displacement current density instead of $\mathbf{q}$, giving

$$\nabla \times \mathbf{H} = \frac{\partial \mathfrak{F}}{\partial t}. \tag{190}$$

This assumption has been completely verified by the experiments of Rowland in America.* If the dielectric is also conducting we retain the term in $\mathbf{q}$, giving

$$\nabla \times \mathbf{H} = 4\pi \mathbf{q} + \frac{\partial \mathfrak{F}}{\partial t}.$$

The term

$$\mathbf{q} + \frac{1}{4\pi}\frac{\partial \mathfrak{F}}{\partial t} \tag{191}$$

is called the *total current* and being equal to a curl is solenoidal, and has no divergence, *i.e.*

$$\nabla \cdot \left(\mathbf{q} + \frac{1}{4\pi}\frac{\partial \mathfrak{F}}{\partial t}\right) = 0.$$

This current therefore moves in closed circuits or paths. It is because of this equation that electricity is said to act like an incompressible fluid. (See § 53.) According to the Electron Theory its compressibility is at most one part in a million.

* Since repeated also by Cremieu and Pender in France.

The complete system of equations for media at rest holding in an *insulating* dielectric are therefore

$$\left.\begin{aligned} \frac{\partial \mathfrak{F}}{\partial t} &= \nabla \times \mathbf{H}, \\ -\frac{\partial \mathfrak{H}}{\partial t} &= \nabla \times \mathbf{F}, \end{aligned}\right\} \quad (192)$$

in combination with

$$\mathfrak{F} = \varepsilon \mathbf{F} \quad \text{and} \quad \mathfrak{H} = \mu \mathbf{H}.$$

ε and μ are the electric and magnetic inductivities respectively, and are defined by these equations. They are constants for a given homogeneous isotropic medium, that is, for non-crystalline media.*

76. Equation of Propagation of Electro-Magnetic Waves. Let us assume that there are no permanent magnets in the space considered, or, in other words, that there is no intrinsic magnetization, symbolically this is expressed by writing

$$\nabla \cdot \mathbf{H} = \operatorname{div} \mathbf{H} = 0.$$

Take the curl of the first of these equations (192),

$$\nabla \times \frac{\partial \mathfrak{F}}{\partial t} = \frac{\partial}{\partial t} \nabla \times \mathfrak{F} = \nabla \times (\nabla \times \mathbf{H}) = -\nabla^2 \mathbf{H} + \nabla(\nabla \cdot \mathbf{H}).$$

Changing from $\mathfrak{F}$ to $\mathbf{F}$ and from $\mathbf{H}$ to $\mathfrak{H}$ and remembering that $\nabla \cdot \mathbf{H} = 0$,

$$\varepsilon \mu \frac{\partial}{\partial t} \nabla \times \mathbf{F} = -\nabla^2 \mathfrak{H}.$$

Differentiating with respect to the time, we obtain

$$\varepsilon \mu \frac{\partial^2}{\partial t^2} (\nabla \times \mathbf{F}) = -\nabla^2 \frac{\partial \mathfrak{H}}{\partial t} = \nabla^2 (\nabla \times \mathbf{F})$$

and also

$$\varepsilon \mu \frac{\partial^2}{\partial t^2} \mathfrak{H} = \nabla^2 \mathfrak{H}. \quad (193)$$

* In order to investigate the form these equations take for crystalline media it is necessary to employ the linear vector-function.

In a similar manner we could show that $\mathbf{F}$, $\mathfrak{F}$, $\mathbf{H}$ and $\mathfrak{H}$ and their curls also satisfy the equation

$$\frac{\partial^2 \phi}{\partial t^2} = a^2 \nabla^2 \phi. \tag{194}$$

This is the differential equation for wave motion, one of the fundamental, partial differential equations of mathematical physics. It can be shown that the velocity of propagation is equal to a, and therefore for an electro-magnetic pulse equal to $\frac{1}{\sqrt{\varepsilon \mu}}$. This turns out to be identical with the velocity of light in vacuo, as it should be if the *ether* is the common medium for the propagation of electrical waves as well as of light. We now believe in fact, that light waves and electrical waves are identical.

77. Poynting's Theorem. Radiant Vector. The energy of the electric field is given by (183) as

$$W_e = \frac{1}{8\pi} \iiint_\infty \mathfrak{F} \cdot \mathbf{F}\, dv,$$

and of the magnetic field, similarly, by

$$W_m = \frac{1}{8\pi} \iiint_\infty \mathfrak{H} \cdot \mathbf{H}\, dv.$$

According to **Joule** the energy due to a current of density $\mathbf{q}$ in the electric field $\mathbf{F}$ is

$$W_J = \iiint_\infty \mathbf{q} \cdot \mathbf{F}\, dv.$$

Let us find the variation of the sum of these three with the time, assuming ε and μ not to vary; then

$$\frac{\partial}{\partial t} \mathfrak{F} \cdot \mathbf{F} = \varepsilon \frac{\partial}{\partial t} \mathbf{F}^2 = 2\varepsilon \frac{\partial \mathbf{F}}{\partial t} \cdot \mathbf{F} = 2\mathbf{F} \cdot \frac{\partial \mathfrak{F}}{\partial t} = 2\mathbf{F} \cdot (\nabla \times \mathbf{H} - 4\pi \mathbf{q}),$$

$$\frac{\partial}{\partial t} \mathfrak{H} \cdot \mathbf{H} = \mu \frac{\partial}{\partial t} \mathbf{H}^2 = 2\mu \frac{\partial \mathbf{H}}{\partial t} \cdot \mathbf{H} = 2\mathbf{H} \cdot \frac{\partial \mathfrak{H}}{\partial t} = -2\mathbf{H} \cdot (\nabla \times \mathbf{F}).*$$

* Heaviside introduces a fictitious magnetic current density $\mathbf{q}_m$ in order to produce a symmetry in the equations. The last parenthesis would then read

$$(\nabla \times \mathbf{F} - 4\pi \mathbf{q}_m).$$

So that the rate change of energy or the *activity* is

$$\frac{\partial W}{\partial t} = \frac{2}{8\pi}\iiint_{\infty} (\mathbf{F}\cdot\nabla\times\mathbf{H} - \mathbf{H}\cdot\nabla\times\mathbf{F})\, dv.$$

By means of (130) and the divergence theorem this may be written

$$\frac{\partial W}{\partial t} = \frac{1}{4\pi}\iiint_{\infty} \nabla\cdot(\mathbf{F}\times\mathbf{H})\, dv = \frac{1}{4\pi}\iint_{S} \mathbf{n}\cdot(\mathbf{F}\times\mathbf{H})dS.$$

Or, in other words, the rate loss of energy per unit volume may be accounted for, by supposing a flux of energy through the bounding surface in unit time per unit surface of amount

$$\mathsf{R} = \frac{\mathbf{F}\times\mathbf{H}}{4\pi}, \tag{195}$$

where R, the energy flux, or *radiant-vector*, is by its form, a vector product, perpendicular both to $\mathbf{F}$ and to $\mathbf{H}$.

78. Magnetic Field Due to a Current. It was proved experimentally by Ampère that the magnetic scalar potential Ω at a point due to a current I whose circuit subtends at the point a solid angle ω is proportional to the product of the current and the solid angle. If we so choose our unit of current that we may write,

$$\Omega = I\omega$$

we thus define a unit called the electro-magnetic unit of current. The magnetic intensity of the field is given, similarly to $\mathbf{F} = -\nabla V$ (106), by

$$\mathbf{H} = -\nabla\Omega = -I\nabla\omega.$$

Its component in any direction $\mathbf{h}$, is by the definition of directional derivative (§ 49)

$$\mathbf{H}\cdot\mathbf{h}_1 = -I\,\mathbf{h}_1\cdot\nabla\omega = -I\frac{\delta\omega}{\delta h}.$$

To find the variation, $\delta\omega$, in the solid angle at a point due to a small displacement $\delta\mathbf{h}$ of the point, notice that it will be the same that will take place, supposing the circuit to move a distance $\delta\mathbf{h}$ in the opposite direction or $-\delta\mathbf{h}$, the point remaining fixed. This motion will cause every element $d\mathbf{r}$ of the circuit to describe a small area $d\mathbf{r}\times\delta\mathbf{h}$ whose component

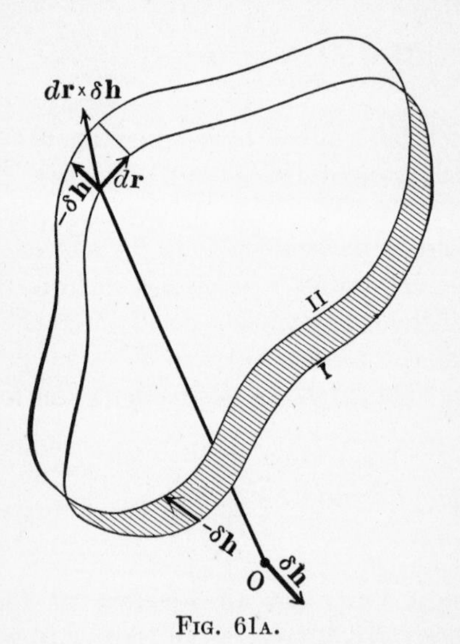

FIG. 61A.

as a vector along $\mathbf{r}$ divided by r^2 is the element of solid angle $d(\delta\omega)$ at the point, due to it. The total change in solid angle $\delta\omega$, due to the motion of the whole circuit, will be the integral of this expression around the contour, thus,

$$d\,\delta\omega = \frac{\mathbf{r}_1\cdot d\mathbf{r}\times\delta\mathbf{h}}{r^2} = \delta\mathbf{h}\cdot\frac{\mathbf{r}_1}{r^2}\times d\mathbf{r} = \delta\mathbf{h}\cdot d\mathbf{r}\times\nabla\frac{1}{r},$$

whose integral around the circuit, after dividing by δh, is

$$\frac{\delta\omega}{\delta h} = \mathbf{h}_1\cdot\int_{\mathfrak{O}} d\mathbf{r}\times\nabla\frac{1}{r},$$

so that
$$\mathbf{H}\cdot\mathbf{h}_1 = -I\,\frac{\partial\omega}{\partial h} = I\,\mathbf{h}_1\cdot\int_{\mathsf{O}}\nabla\frac{1}{r}\times d\mathbf{r}. \tag{196}$$

Since this is true for any direction $\mathbf{h}_1$, we may write that the element of magnetic intensity $d\mathsf{H}$, at a point, due to the element $d\mathbf{r}$ of the circuit, is

$$d\mathsf{H} = I\left(\nabla\frac{1}{r}\right)\times d\mathbf{r} = -\frac{I\,\mathbf{r}_1\times d\mathbf{r}}{r^2}. \tag{197}$$

This expression is determined to any function of r *près* which, when integrated around the circuit, vanishes. So that to this extent it is arbitrary. The equation shows that the force due to an element is perpendicular to the element and to the radius vector. The radius $\mathbf{r}$ being drawn from the point to the circuit, and the current being positive in the direction of $d\mathbf{r}$, the order of the factors is taken so as to give the right direction to H. This is the familiar expression for the magnetic intensity at O due to $d\mathbf{r}$.

$$d\mathsf{H} = \frac{I\,dr\sin\theta}{r^2}\ \perp \text{ to element and to } r.$$

79. Mechanical Force on an Element of Circuit. The magnetic intensity $d\mathsf{H}$, above, is the force with which a unit positive pole placed at the point would be acted upon in the field due to the current I in $d\mathbf{r}$. By the principle of equal action and reaction the element of circuit would be acted upon by this amount but in the opposite direction. The force on the element $d\mathbf{r}$ due to unit pole at the origin is therefore

$$d\mathsf{F} = I\,d\mathbf{r}\times\frac{\mathbf{r}_1}{r^2} = I\,d\mathbf{r}\times\boldsymbol{\phi}. \tag{198}$$

So that the force on an element of current is proportional to the current strength I, to the length of the element $d\mathbf{r}$, and to the strength of the field at the element $\boldsymbol{\phi}$. The factor of proportionality is the sine of the angle between $d\mathbf{r}$ and $\boldsymbol{\phi}$ and the force is at right angles to their plane.

Hence the force on the elementary current $I'd\mathbf{r}'$ in a field due to another elementary current $I''d\mathbf{r}''$ at O whose field at the element $d\mathbf{r}'$ is by (197) and (198)

$$I'' \frac{d\mathbf{r}'' \times \mathbf{r}_1}{r^2},$$

where the order of the vector product is reversed to take into account the change of direction in $\mathbf{r}$, or

$$d^2\mathbf{F} = \frac{(I'd\mathbf{r}') \times (I''d\mathbf{r}'' \times \mathbf{r}_1)}{r^2} = \frac{I'I''d\mathbf{r}' \times (d\mathbf{r}'' \times \mathbf{r}_1)}{r^2}. \tag{199}$$

We may resolve this expression immediately into components along the radius vector $\mathbf{r}$ and along the element $d\mathbf{r}$ by an expansion of the triple vector product, giving

$$d^2F = \frac{I'I''}{r^2}\, dr'dr'' \left[\mathbf{r}_1 \cos(dr' \cdot dr'') - d\mathbf{r}'' \cos(dr'r) \right]. \tag{200}$$

The X component is

$$d^2\mathbf{F}_x = \frac{I'I''}{r^2}\, dr'dr'' \Big[\cos(rx) \cos(dr'dr'') - \cos(dr''x) \cos(dr'r) \Big], \text{ etc.}$$

These are well-known results.

80. Theorem on the Line Integral of the Normal Component of a Vector Around a Closed Circuit. By means of Stokes' Theorem the following useful transformation analogous to it may be proved:

$$\int_{\mathfrak{O}} \mathbf{q} \times d\mathbf{r} = \iint_{\text{cap}} [(\nabla \cdot \mathbf{q})\, \mathbf{n} - \nabla (\mathbf{q} \cdot \mathbf{n})]\, dS. \tag{201}$$

$\mathbf{q}$ is any vector function of $\mathbf{r}$, and $\mathbf{n}$ is the unit normal to the element of surface dS. In this case the line integral of the *normal* component of the vector $\mathbf{q}$ along a closed circuit is taken, instead of the more usual, tangential component.

The result is a vector one, and it is shown to be expressible as the surface integral of a certain other vector quantity related to **q**, taken over any cap which is bounded by the

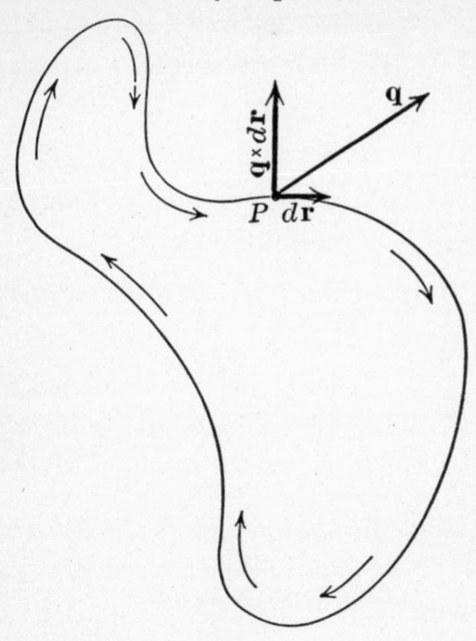

Fig. 62.

circuit. Let **H** be this line integral, and form its scalar product with the arbitrary *constant* vector **c**,

$$\mathbf{c}\cdot\mathbf{H} = \mathbf{c}\cdot\int_{\circlearrowleft}\mathbf{q}\times d\mathbf{r} = \int_{\circlearrowleft}\mathbf{c}\cdot\mathbf{q}\times d\mathbf{r} = \int_{\circlearrowleft} d\mathbf{r}\cdot(\mathbf{c}\times\mathbf{q}).$$

This last expression being in the form of a tangential line integral may be transformed by Stokes' Theorem (136) into

$$\mathbf{c}\cdot\mathbf{H} = \iint_{\text{cap}} [\mathbf{n}\cdot\nabla_q\times(\mathbf{c}\times\mathbf{q})]\, dS,$$

where the ∇ differentiates **q** alone. Now

$$\nabla_q\times(\mathbf{c}\times\mathbf{q}) = \mathbf{c}\,(\nabla\cdot\mathbf{q}) - (\mathbf{c}\cdot\nabla)\,\mathbf{q},$$

so that

$$\mathbf{c}\cdot\mathbf{H} = \iint_{\text{cap}} [\mathbf{c}\cdot\mathbf{n}\,(\nabla\cdot\mathbf{q}) - \mathbf{n}\cdot(\mathbf{c}\cdot\nabla)\,\mathbf{q}]\, dS.$$

But (117)

$$(\mathbf{c}\cdot\nabla_q)(\mathbf{q}\cdot\mathbf{n}) = \mathbf{n}\cdot(\mathbf{c}\cdot\nabla)\,\mathbf{q}.$$

The differentiation refers to $\mathbf{q}$ alone, as in the above integral the *properties of the region are independent of the surface considered*, and we may substitute $(\mathbf{c}\cdot\nabla_q)(\mathbf{q}\cdot\mathbf{n})$ for the second term in the integral, hence,

$$\mathbf{c}\cdot\mathbf{H} = \mathbf{c}\cdot\int\!\!\int_{\text{cap}} [\mathbf{n}\nabla_q\cdot\mathbf{q} - \nabla_q(\mathbf{q}\cdot\mathbf{n})]\,dS,$$

and finally, since $\mathbf{c}$ is an *arbitrary* vector,

$$\mathbf{H} = \pm\int_{\circlearrowleft}\mathbf{q}\times d\mathbf{r} = \pm\int\!\!\int_{\text{cap}} [\mathbf{n}\nabla\cdot\mathbf{q} - \nabla(\mathbf{q}\cdot\mathbf{n})]\,dS, \quad (202)$$

which sign to take, depends upon the direction of integration around the contour. This theorem is originally due to Tait and to McAulay, who gave it in a much more general form, including Stokes' and other theorems, as special cases.

81. Expression for the Field at any Point in Space Due to a Current. We may use this theorem to transform the integral, giving the magnetic force $\mathbf{H}$ at any point in space due to a current of electricity in a closed circuit.

If the magnetic potential is Ω, I the current in the circuit, and ω the solid angle subtended by the circuit at the point, we have, by definition,

$$\mathbf{H} = -\nabla\Omega = -I\nabla\omega = -I\nabla\int\!\!\int_{\text{cap}} \frac{\mathbf{n}\cdot\mathbf{r}_1}{\mathbf{r}^2}\,dS,$$

the surface integral being taken over any surface with the circuit for bounding edge, but not passing through the point.

Now
$$\nabla\int\!\!\int_{\text{cap}} \frac{\mathbf{n}\cdot\mathbf{r}_1}{\mathbf{r}^2}\,dS = -\nabla\int\!\!\int_{\text{cap}} \mathbf{n}\cdot\nabla\frac{1}{r}\,dS$$
$$= -\int\!\!\int_{\text{cap}} \nabla\left(\mathbf{n}\cdot\nabla\frac{1}{r}\right)dS.$$

Employing the above theorem (202) and remembering that

$$\nabla\cdot\nabla\frac{1}{r} = \nabla^2\frac{1}{r} = 0, \quad \text{by (148)},$$

$$\mathbf{H} = I \int\int_{\text{cap}} \nabla\left(\mathbf{n}\cdot\nabla \frac{1}{r}\right) dS = -I \int_{\mathfrak{C}} \nabla \frac{1}{r} \times d\mathbf{r}$$

$$= \pm I \int_{\mathfrak{C}} d\mathbf{r} \times \nabla \frac{1}{r}, \tag{203}$$

which result is in agreement with (197). See also (202).

We may thus write

$$d\mathbf{H} = I\, d\mathbf{r} \times \nabla \frac{1}{r} + \boldsymbol{\phi},$$

in Cartesian

$$dH_x = \frac{I}{r^3} \{ dy_1 (z - z_1) - dz_1 (y - y_1) \}, \text{ etc.}$$

where x_1, y_1, z_1, are the coördinates of a point in the circuit, x, y, z those of the point P, so that the small magnetic field $d\mathbf{H}$ due to an element $d\mathbf{r}$ is determined to a function $\boldsymbol{\phi}$ *près* such that when integrated around a closed circuit the result vanishes. So to a certain extent this resolution of the field is artificial, and may or may not be the correct one, but in any case this as well as any other possible resolution will give the correct value for $\mathbf{H}$ above *when integrated* around a *closed circuit;* we have but very scanty knowledge of the fields due to unclosed circuits.

82. Mutual Energy of Two Circuits. Inductance. Neumann's Integral. Consider two circuits carrying currents I' and I''. The mechanical force on one of them due to the field of the other is the integral of equation (199) taken once around each circuit. Since

$$\nabla \frac{1}{r} = -\frac{\mathbf{r}_1}{r^2},$$

we may write

$$\mathbf{F} = I'I'' \int_{\mathfrak{C}'}\int_{\mathfrak{C}''} d\mathbf{r}' \times \left(\nabla \frac{1}{r} \times d\mathbf{r}''\right)$$

$$= I'I'' \int_1\int_2 \left\{ \nabla \frac{1}{r} (d\mathbf{r}' \cdot d\mathbf{r}'') - d\mathbf{r}'' \left(d\mathbf{r}' \cdot \nabla \frac{1}{r} \right) \right\}.$$

Let one circuit be now displaced in any arbitrary manner, so that any point on it is moved a distance $\delta \mathbf{r}'$. The work done in this displacement is

$$\mathbf{F}\cdot\delta\mathbf{r}' = I'I''\int_1\int_2 \left\{ \left(\delta\mathbf{r}'\cdot\nabla\frac{1}{r}\right)(d\mathbf{r}'\cdot d\mathbf{r}'') - (\delta\mathbf{r}'\cdot d\mathbf{r}'')\left(d\mathbf{r}'\cdot\nabla\frac{1}{r}\right)\right\}.$$

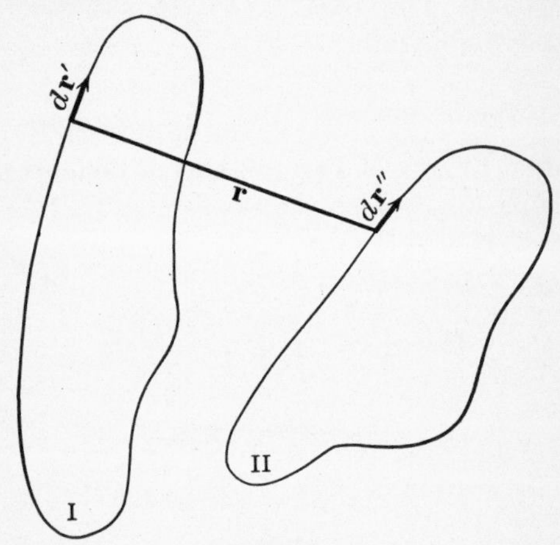

FIG. 63.

Integrating the second term by parts and remembering that $d\,\delta\mathbf{r} = \delta\,d\mathbf{r}$,

$$\int_1 (\delta\mathbf{r}'\cdot d\mathbf{r}'')\left(d\mathbf{r}'\cdot\nabla\frac{1}{r}\right) = \frac{1}{r}\,\delta\mathbf{r}'\cdot d\mathbf{r}''\Big/ - \int_1 \frac{1}{r}\,d\,(\delta\mathbf{r}'\cdot d\mathbf{r}''),$$

the integrated portion vanishes for a closed circuit, hence

$$\begin{aligned}\mathbf{F}\cdot\delta\mathbf{r}' &= I'I''\int_1\int_2 \left\{ (d\mathbf{r}'\cdot d\mathbf{r}'')\left(\delta\frac{1}{r}\right) + \left(\frac{1}{r}\right)\delta\,(d\mathbf{r}'\cdot d\mathbf{r}'')\right\} \\ &= I'I''\delta_r\int_1\int_2 \frac{d\mathbf{r}'\cdot d\mathbf{r}''}{r}.\end{aligned}$$

As the assumed motion of the circuit is arbitrary we may then find the force in any direction by finding the change in

$$I'I''\int_1\int_2 \frac{d\mathbf{r}'\cdot d\mathbf{r}''}{r}. \tag{204}$$

This integral due to Neumann represents the mutual energy of the two circuits. When the currents I' and I'' are each unity, the integral gives the **mutual-inductance** of the circuits. When taken twice around a single circuit it gives the **self-inductance** of the circuit. It is sometimes called the Electro-Dynamic Potential, as by its variation we obtain the electro-dynamic forces.

Vector Potential Due to a Current.

83. Mutual Energy of Two Systems of Conductors. If the magnetic force due to a current be denoted by **H**, a solenoidal vector, *i.e.*, $\nabla \cdot \mathbf{H} = 0$, we may write, by means of the theorem of Helmholtz (175),

$$\mathbf{H} = \frac{1}{4\pi} \nabla \times \iiint_{\infty} \frac{\nabla \times \mathbf{H}}{r} dv.$$

But (188)

$$\nabla \times \mathbf{H} = 4\pi \mathbf{q},$$

where **q** is the current density, so that substituting, we obtain

$$\left. \begin{aligned} \mathbf{H} &= \nabla \times \iiint_{\infty} \frac{\mathbf{q}}{r} dv = \nabla \times \mathbf{Q}, \\ \text{where} \quad & \\ \mathbf{Q} &\equiv \iiint_{\infty} \frac{\mathbf{q}}{r} dv. \end{aligned} \right\} \qquad (205)$$

Q is called the potential due to the current distribution **q**, or the *vector-potential* belonging to the magnetic force **H**. The word *potential* is used because it is formed in a manner analogous to the potential due to a scalar distribution of matter ρ. Notice also that the force vector **H** is obtained from *the vector* **Q** in a manner *analogous* to the way the force vector **F** is obtained from the scalar V, where

$$V \equiv \iiint_{\infty} \frac{\rho}{r} dv \quad \text{and} \quad \mathbf{F} = \nabla V.$$

hence the name vector-potential.

We may now transform the magnetic energy in terms of the field $\mathbf{H}$,

$$W_m = \frac{1}{8\pi}\iiint_\infty \mathbf{H}^2 dv = \frac{1}{8\pi}\iiint_\infty \mathbf{H}\cdot\mathbf{H}\, dv,$$

into

$$W_m = \frac{1}{8\pi}\iiint_\infty \mathbf{H}\cdot\nabla\times\mathbf{Q}\, dv. \tag{206}$$

Integration Theorem. In general (130) we have, where $\mathbf{H}$ and $\mathbf{Q}$ are any two vectors,

$$\nabla\cdot(\mathbf{H}\times\mathbf{Q}) = \mathbf{Q}\cdot\nabla\times\mathbf{H} - \mathbf{H}\cdot\nabla\times\mathbf{Q},$$

the minus sign belonging to the term in which the cyclical order has been changed. Integrating over all space and using the divergence theorem (121), S being the bounding surface,

$$\iiint_\infty \nabla\cdot(\mathbf{H}\times\mathbf{Q})\, dv = \iiint_\infty (\mathbf{Q}\cdot\nabla\times\mathbf{H} - \mathbf{H}\cdot\nabla\times\mathbf{Q})\, dv$$
$$= \iint_S \mathbf{n}\cdot(\mathbf{H}\times\mathbf{Q})\, dS.$$

Substituting (206) in this equation and remembering that the surface integral vanishes at infinity, because there the magnetic force vanishes, and also that $\nabla\times\mathbf{H} = 4\pi\mathbf{q}$ there remains

$$W_m = \frac{1}{8\pi}\iiint_\infty \mathbf{Q}\cdot\nabla\times\mathbf{H}\, dv$$
$$= \frac{1}{2}\iiint_\infty \mathbf{Q}\cdot\mathbf{q}\, dv.$$

Now replacing $\mathbf{Q}$ by its value, we have finally

$$W_m = \frac{1}{2}\iiint_\infty \iiint_\infty \frac{\mathbf{q}\cdot\mathbf{q}'}{r}\, dv\, dv'. \tag{207}$$

This sextuple integral covers the whole of space *twice*.

Let the only portions of space having any current density be two closed circuits. The wire forming the circuits may have a small but finite cross-section.

Place $\mathbf{q}\,dv \equiv I' d\mathbf{r}' \qquad \mathbf{q}' dv' \equiv I'' d\mathbf{r}''$,

where I' and I'' are called the currents in the two circuits, respectively. The integrals then reduce to a double line integral each integral to be taken once around each of the circuits, so that

$$W_m = \frac{I' I''}{2} \int_{\text{C}} \int_{\text{C}} \frac{d\mathbf{r}' \cdot d\mathbf{r}''}{r}. \tag{208}$$

This expression is really identical with Neumann's Integral (204). The factor $\frac{1}{2}$ is due to the fact that by the convention in (207) the integrals cover each of the circuits twice and hence would give twice the value of (204).

84. Mutual and Self-Energies of Two Circuits. *Each* integral being taken over *both* circuits, (208) may be broken up into four parts,

$$W_m = \frac{I_1^2}{2} \int_1 \int_1 \frac{d\mathbf{r}' \cdot d\mathbf{r}''}{r} + \frac{I_1 I_2}{2} \int_1 \int_2 \frac{d\mathbf{r}' \cdot d\mathbf{r}''}{r}$$

$$+ \frac{I_2 I_1}{2} \int_2 \int_1 \frac{d\mathbf{r}' \cdot d\mathbf{r}''}{r} + \frac{I_2^2}{2} \int_2 \int_2 \frac{d\mathbf{r}' \cdot d\mathbf{r}''}{r}.$$

The second and third parts are evidently equal, so that we may write for their sum

$$I_1 I_2 \int_1 \int_2 \frac{d\mathbf{r}' \cdot d\mathbf{r}''}{r},$$

where here, each integral is taken around its corresponding circuit once.

If we call the integrals

$$L_1 = \int_1\int_1 \frac{d\mathbf{r}'\cdot d\mathbf{r}''}{r}, \quad L_2 = \int_2\int_2 \frac{d\mathbf{r}'\cdot d\mathbf{r}''}{r},$$

$$M_{12} = \int_1\int_2 \frac{d\mathbf{r}'\cdot d\mathbf{r}''}{r},$$

we may write for the magnetic energy of the field due to both currents

$$W_m = \tfrac{1}{2} L_1 I_1^2 + M_{12} I_1 I_2 + \tfrac{1}{2} L_2 I_2^2. \tag{209}$$

The integrals L_1 or L_2 and M_{12} are called the self-inductances and mutual-inductance of the circuits respectively.

EXERCISES AND PROBLEMS.

1. If the line integral of the forces in any field around a closed contour is zero for any such contour, the forces in the field form a conservative system.

2. Show that the surface integral of a *scalar* point-function V taken over any closed surface is equal to the volume integral of its grad (∇V) taken throughout the volume of that surface; that is,

$$\int\int_S V\,\mathbf{n}\,dS = -\int\int\int_{\text{vol}} \nabla V\,dv.$$

3. Show that the line integral of a scalar point-function, V, around a closed contour is equal to the surface integral of the vector product of the normal by its gradient taken over any cap to the contour; that is, prove

$$\int_{\circlearrowleft} V\,d\mathbf{r} = \int\int_{\text{cap}} \mathbf{n}\times\nabla V\,dS.$$

4. Using the divergence theorem, let $\mathbf{q} = \mathbf{r}_1$, and prove that the potential of a body may be represented by the surface integral

$$V = \frac{1}{2}\int\int_s \mathbf{n}\cdot\mathbf{r}_1\,\rho\,dS - \frac{1}{2}\int\int\int_{\text{vol}} \mathbf{r}_1\cdot\nabla\rho\,dv.$$

5. If Poisson's equation holds,

i.e., $$\nabla^2 V = 4\,\pi\rho = \text{div}\,\mathbf{F} \text{ and if } \nabla\rho = 0$$

show that the potential of a body in the last example becomes

$$V = \frac{1}{8\,\pi}\int\!\!\int_{\text{surf}} \mathbf{r}_1 \cdot \mathbf{n} \nabla^2 V\, dS = \frac{1}{8\,\pi}\int\!\!\int \mathbf{r}_1 \cdot \mathbf{n}\ \text{div}\ \mathbf{F}\, dS,$$

so that if the force at every point of S be known it is possible to compute the potential.

6. By drawing a small cylindrical box enclosing a portion dS of a surface charged with a surface density of electricity σ, and making the cylindrical sides everywhere parallel to the lines of force, show that there is a change in the *normal* component of the flux of moment $4\,\pi\sigma$.

7. The curl of the curl of a solenoidal vector such that the three functions which give the strengths of its components parallel to **i**, **j** and **k** satisfy Laplace's Equation, vanishes.

8. If the lines of a vector, F, are all parallel to a plane and the vector has the same value at all points in any line perpendicular to the plane, the vector is perpendicular to its curl,

i.e., $$\mathbf{F} \cdot \nabla \times \mathbf{F} = 0.$$

9. Compare the results of the last problem with those of § 60. Can you devise any other functions, the lines of which are everywhere perpendicular to its curl?

10. If the lines of a vector are circles parallel to the **ij**-plane with centers on the **k** axis, and if the intensity of the vector is a function $\mathbf{f}(\mathbf{r})$ of the distance from the **k** axis, a vector everywhere parallel to the **k** axis, of intensity $\mathbf{F}(\mathbf{r})$, where $\mathbf{f}(\mathbf{r}) = -\dfrac{d\mathbf{F}}{dr}$ is a vector potential-function of the original vector. Is the original vector solenoidal?

CHAPTER VII.

APPLICATIONS TO DYNAMICS, MECHANICS, AND HYDRODYNAMICS.

Equations of Motion of a Rigid Body.

85. Equations for Translation. D'Alembert's Principle, upon which **Lagrange** founded the whole subject of analytical mechanics, may be written

$$\sum\left(m\frac{d^2\mathbf{r}}{dt^2}-\mathbf{F}\right).\delta\mathbf{r}=0, \tag{210}$$

where $\delta\mathbf{r}$ is any possible arbitrary or virtual displacement compatible with the constraints imposed upon the system, and where the Σ sums for all the particles.

In order to deduce the equations of motion of translation assume the virtual displacement to be the *same* for all points of the system, as this is the definition of pure translatory motion. It then follows, since we may now take the $\delta\mathbf{r}$ from under Σ sign, that

$$\delta\mathbf{r}\cdot\sum\left(m\frac{d^2\mathbf{r}}{dt^2}-\mathbf{F}\right)=0,$$

and since $\delta\mathbf{r}$ is arbitrary, that

$$\sum\left(m\frac{d^2\mathbf{r}}{dt^2}-\mathbf{F}\right)=0, \tag{211}$$

which are the ordinary equations:

$$\sum\left(m\frac{d^2x}{dt^2}-X\right)=0.$$

$$\sum\left(m\frac{d^2y}{dt^2}-Y\right)=0.$$

$$\sum\left(m\frac{d^2z}{dt^2}-Z\right)=0.$$

See papers by Ziwet and Field in American Mathematical Monthly, 1914, pp. 105–113 and by Rees same journal 1923, pp. 290–296 for interesting vectorial treatments of kinematics and motion of rigid bodies.

Motion of Center of Mass. Let $\bar{\mathbf{r}}$ be the vector to the center of mass or centroid of the system; then, by the definition of this point for which (20)

$$\bar{\mathbf{r}} \sum m = \sum m\, \mathbf{r},$$

we have by differentiation

$$\frac{d^2\bar{\mathbf{r}}}{dt^2} \sum m = \sum m \frac{d^2\mathbf{r}}{dt^2},$$

so that finally equation (211) may be written

$$\frac{d^2\bar{\mathbf{r}}}{dt^2} \sum m = \sum \mathbf{F}, \qquad (212)$$

or, in words, the motion of translation of the centroid of a system of bodies moves precisely as if all the forces of the system were applied to the total mass concentrated at that point. This reduces the problem of the translatory motion of the *system* to that of the motion of a *single point*. An interesting example of this property is seen in the case of the motion of a shell which explodes while describing its path in space. As the resultant of the actions and the reactions which are produced when the shell explodes is zero, the path of the center of mass of the fragments is the identical parabola the center of mass of the shell would have described had it not exploded. In other words, the path of the center of mass remains unchanged by the explosion. The center of mass of a thrown stick describes a smooth parabola, as it whirls through the air.

The kinetic energy of translation of the body is evidently given by

$$T = \frac{1}{2} \sum m \left(\frac{d\mathbf{r}}{dt}\right)^2 = \frac{1}{2} M \bar{\mathbf{q}}^2, \qquad (213)$$

where $\bar{\mathbf{q}}$ is the velocity of the center of mass of the system, where $M = \Sigma m$ is its total mass and because all points of the system have the same velocity.

86. Equations for Rotation. To deduce the equations of motion for rotation, let $\delta\boldsymbol{\omega}$ be an elementary rotation, then

$$\delta\mathbf{r}_s = \delta\boldsymbol{\omega}\times\mathbf{r}_s,$$

where $\delta\mathbf{r}_s$ is an arbitrary possible infinitesimal motion due to the rotation of any particle m_s of the system about some axis, $\boldsymbol{\omega}$. Substituting this in d'Alembert's equation, we obtain

$$\sum m\frac{d^2\mathbf{r}}{dt^2}\cdot\delta\boldsymbol{\omega}\times\mathbf{r} = \sum \mathbf{F}\cdot\delta\boldsymbol{\omega}\times\mathbf{r},$$

and with obvious transformations

$$\sum m\,\delta\boldsymbol{\omega}\cdot\mathbf{r}\times\frac{d^2\mathbf{r}}{dt^2} = \sum \delta\boldsymbol{\omega}\cdot\mathbf{r}\times\mathbf{F}.$$

We shall now assume that the particles of the system rotate about the same axis, so that $\delta\boldsymbol{\omega}$ shall be the same for all the particles and may be divided out, and remembering that (34)

$$\frac{d\mathbf{r}}{dt}\times\frac{d\mathbf{r}}{dt} \equiv 0,$$

we obtain
$$\sum m\,\mathbf{r}\times\frac{d^2\mathbf{r}}{dt^2} = \frac{d}{dt}\sum m\,\mathbf{r}\times\frac{d\mathbf{r}}{dt} = \sum \mathbf{r}\times\mathbf{F} \qquad (214)$$

for the equation of motion of *rotation* of a system about an axis. The motion of a *rigid* system is of course a special case of this. This equation expands into the familiar Cartesian ones,

$$\left.\begin{aligned}
\frac{d}{dt}\sum m\left(y\frac{dz}{dt} - z\frac{dy}{dt}\right) &= \sum (yZ - zY),\\
\frac{d}{dt}\sum m\left(z\frac{dx}{dt} - x\frac{dz}{dt}\right) &= \sum (zX - xZ),\\
\frac{d}{dt}\sum m\left(x\frac{dy}{dt} - y\frac{dx}{dt}\right) &= \sum (xY - yX),
\end{aligned}\right\} \qquad (215)$$

about the three rectangular axes, by the ordinary rules.

Defining
$$\mathbf{M} \equiv \Sigma\, \mathbf{r}\times\mathbf{F} \qquad (216)$$

as the **moment** *of the applied forces* about the axis of rotation, $\boldsymbol{\omega}$, and

$$\mathbf{H} \equiv \sum m\, \mathbf{r} \times \frac{d\mathbf{r}}{dt} \tag{217}$$

as the **moment of momentum** about the same axis, the above equation (214) may be written

$$\frac{d}{dt}\mathbf{H} = \mathbf{M}, \tag{218}$$

or even

$$\dot{\mathbf{H}} = \mathbf{M},$$

or, in words, the rate of increase of **angular momentum** † about the axis of rotation of a system is equal to the moment of the impressed forces about that same axis.

Kinetic Energy of Rotation. Moment of Inertia.* The kinetic energy of rotation of the system rotating with angular velocity $\boldsymbol{\omega}$ is (44)

$$T = \tfrac{1}{2}\, \Sigma m\, \mathbf{q}^2 = \tfrac{1}{2}\, \Sigma m\, (\boldsymbol{\omega} \times \mathbf{r})^2.$$

If the system moves as a rigid body all of the $\boldsymbol{\omega}$'s are the same, so that

$$T = \tfrac{1}{2}\, \boldsymbol{\omega}^2 \Sigma m\, (\boldsymbol{\omega}_1 \times \mathbf{r})^2. \tag{219}$$

But $(\boldsymbol{\omega}_1 \times \mathbf{r})^2$ is the perpendicular squared from the point $\mathbf{r}$ to the axis of rotation $\boldsymbol{\omega}$ (AP in **Fig. 31**), so that the expression

$$\Sigma m\, (\boldsymbol{\omega}_1 \times \mathbf{r})^2$$

means that every elementary mass is to be multiplied by the square of its distance from the axis of rotation and that their sum is to be taken. This quantity is called the *Moment of Inertia** of the system about the axis $\boldsymbol{\omega}_1$; it evidently varies with the direction of $\boldsymbol{\omega}$. We may then define the moment of inertia, I, about an axis $\boldsymbol{\omega}$, by the equation

$$I_\omega \equiv \Sigma m\, (\boldsymbol{\omega}_1 \times \mathbf{r})^2 = M k_\omega^{\,2}, \tag{220}$$

where k_ω, also defined by the above equation, is called the **Radius of Gyration** about the axis $\boldsymbol{\omega}$. The radius of gyration is, therefore, the distance from the axis of rotation at

* Or, better, Rotational Mass.

† Same as Moment of Momentum.

which, if the total mass M of the system were placed, its moment of inertia would remain unchanged.

The *total* kinetic energy of a rigid system moving in any manner may then be written

$$T = \frac{1}{2} M \overline{\mathbf{q}}^2 + \frac{1}{2} I_\omega \boldsymbol{\omega}^2. \tag{221}$$

M, the mass of the body, is an absolute constant,* but I_ω, as stated above, varies with the direction of the axis about which the system rotates, and hence the treatment of rotation is essentially more complicated than that of pure translation.

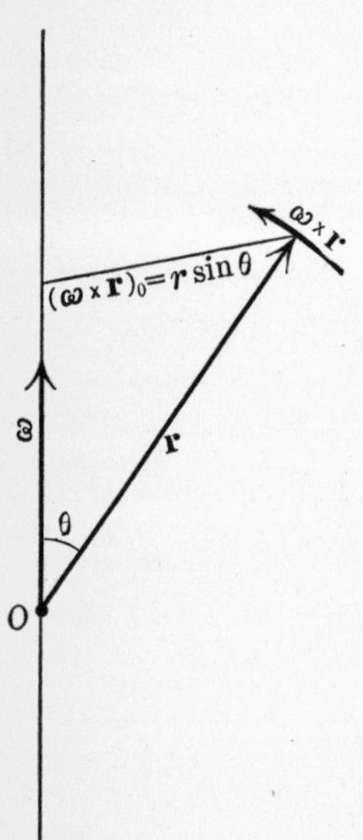

Fig. 64.

We shall treat of the motion of rotation more in detail, not only for its intrinsic interest, but also because it introduces naturally the Linear Vector-Function and some of its elementary properties.

87. Linear Vector-Function. Instantaneous Axis. Consider a rigid body of mass M rotating in any manner about a fixed point. This precludes any translatory motion of the body which is now one of pure rotation at any instant about some axis necessarily passing through this fixed point. This

* In the Electron Theory of Matter the inertia of a particle, at least in part, is accounted for by the electrical charge which we know the particle carries. The resistance of an electrical charge to acceleration is not constant but is a function of the velocity, and theoretically becomes infinite when its velocity approaches that of light. The apparent inertia of such a particle is therefore not constant. But for any velocities with which we are likely to deal in mechanical systems these variations in inertia are inappreciable. The ordinary equations of mechanics are then first approximations only, but for ordinary velocities, up to 10,000 km. per sec. say, are extremely close to the truth.

axis, which may vary continuously in direction, is called the Instantaneous Axis of rotation. Let the angular velocity about this axis at any instant be represented by a vector of length $\boldsymbol{\omega} = \mathbf{f}(t)$ in the direction of it and in the conventional sense, *i.e.*, that of the motion of progression and direction of rotation of a corkscrew.

As the velocity of any point $\mathbf{r}$ is

$$\mathbf{q} = \frac{d\mathbf{r}}{dt} = \boldsymbol{\omega}\times\mathbf{r},$$

we may write for the moment of momentum $\mathbf{H}$

$$\mathbf{H} = \sum m\, \mathbf{r}\times\frac{d\mathbf{r}}{dt}$$

$$= \sum m\, \mathbf{r}\times(\boldsymbol{\omega}\times\mathbf{r}) = \sum m(\boldsymbol{\omega}\, \mathbf{r}^2 - \mathbf{r}\, \boldsymbol{\omega}\cdot\mathbf{r}); \qquad (222)$$

thus $\mathbf{H}$ is a vector-function linear in $\boldsymbol{\omega}$, $\boldsymbol{\phi\omega}$ say.

This particular function $\boldsymbol{\phi\omega}$ has a number of important properties which are evident upon inspection. If $\boldsymbol{\tau}$ and $\boldsymbol{\sigma}$ are any two vectors, the following equations hold:

$$\left.\begin{array}{lll} & \boldsymbol{\phi}(\boldsymbol{\tau} \pm \boldsymbol{\sigma}) = \boldsymbol{\phi}\boldsymbol{\tau} \pm \boldsymbol{\phi}\boldsymbol{\sigma}, & (a) \\ a = \text{const.} & \boldsymbol{\phi} a\boldsymbol{\tau} = a\boldsymbol{\phi}\boldsymbol{\tau}, & (b) \\ & d(\boldsymbol{\phi}\boldsymbol{\tau}) = \boldsymbol{\phi}\, d\boldsymbol{\tau}, & (c) \\ \text{and} & \boldsymbol{\tau}\cdot\boldsymbol{\phi}\boldsymbol{\sigma} = \boldsymbol{\sigma}\cdot\boldsymbol{\phi}\boldsymbol{\tau}. & (d) \end{array}\right\} \qquad (223)$$

In particular when a linear vector-function has the property represented by (223) (*d*) it is said to be **Self-Conjugate.**

Form the scalar product of $\mathbf{H}$ and $\boldsymbol{\omega}$.

$$\boldsymbol{\omega}\cdot\mathbf{H} = \boldsymbol{\omega}\cdot\boldsymbol{\phi}\,\boldsymbol{\omega} = \Sigma m\, \boldsymbol{\omega}\cdot(\mathbf{r}\times(\boldsymbol{\omega}\times\mathbf{r}))$$

$$= \Sigma m\, (\boldsymbol{\omega}\times\mathbf{r})\cdot(\boldsymbol{\omega}\times\mathbf{r}) = \Sigma m\, (\boldsymbol{\omega}\times\mathbf{r})^2$$

$$= \boldsymbol{\omega}^2 I_{\omega} = 2\,T \qquad \text{by (221).} \qquad \mathbf{(224)}$$

Now, since $\phi\omega$ is linear in ω, the scalar product $\omega \cdot \phi\omega$ is a *quadratic scalar-function* of ω, and hence represents when equated to a constant, a quadric surface.

88. Motion under No Forces. Invariable Plane. Assuming no applied forces,

$$\frac{d}{dt}\mathrm{H} = 0,$$

so that

$$\mathrm{H} = \text{const. vector},$$

or, in words, under no applied forces the moment of momentum of a rigid system remains constant in magnitude and direction. It remains perpendicular to the plane, called Invariable Plane, whose equation is

$$\mathbf{r} \cdot \mathrm{H} = \text{const.}$$

Also, since the energy (kinetic) of the system is conserved, it follows that the moment of inertia I_ω about any direction ω is inversely proportional to the square of the radius vector in the quadric $\omega \cdot \phi\omega = \text{const.}$, because

$$\omega \cdot \mathrm{H} = \omega \cdot \phi\omega = 2\,T = \omega^2 I_\omega,$$

so that

$$I_\omega = \frac{2\,T}{\omega^2} = \frac{\text{const.}}{\omega^2} \tag{225}$$

This equation also says that with a given amount of energy the body rotates the faster the smaller I_ω is; *i.e.*,

$$\omega \propto \frac{1}{\sqrt{I_\omega}}.$$

Poinsot Ellipsoid. Since evidently no finite body has an infinite or a zero moment of inertia about any axis, the quadric surface $\omega \cdot \phi\omega = \text{const.}$ must be one the radius vector of which has a finite minimum and maximum value; that is, it must be an ellipsoid. This ellipsoid is called the **Momental** or Poinsot Ellipsoid. Let us consider this surface more in detail and incidentally show its expansion in Cartesian form.

If H_1, H_2, H_3 are the components of **H** about the three axes **i**, **j**, and **k**, then (222)

$$\begin{aligned}\mathbf{H} \equiv \boldsymbol{\phi}\boldsymbol{\omega} &= \sum m(\boldsymbol{\omega}\, \mathbf{r}^2 - \mathbf{r}\, \boldsymbol{\omega}\cdot\mathbf{r}) = H_1\mathbf{i} + H_2\mathbf{j} + H_3\mathbf{k} \\ &= \{\Sigma m\omega_1(x^2+y^2+z^2) - \Sigma mx(\omega_1 x+\omega_2 y+\omega_3 z)\}\, \mathbf{i} \\ &+ \{\Sigma m\omega_2(x^2+y^2+z^2) - \Sigma my(\omega_1 x+\omega_2 y+\omega_3 z)\}\, \mathbf{j} \\ &+ \{\Sigma m\omega_3(x^2+y^2+z^2) - \Sigma mz(\omega_1 x+\omega_2 y+\omega_3 z)\}\, \mathbf{k} \\ &= \{\omega_1 \Sigma m(y^2+z^2) - \omega_2 \Sigma m\, xy \quad - \omega_3 \Sigma m\, xz\}\, \mathbf{i} \\ &+ \{-\omega_1 \Sigma m\, yx \quad + \omega_2 \Sigma m(z^2+x^2) - \omega_3 \Sigma m\, yz\}\, \mathbf{j} \qquad (226) \\ &+ \{-\omega_1 \Sigma m\, xz \quad - \omega_2 \Sigma m\, yz \quad + \omega_3 \Sigma m(x^2+y^2)\}\, \mathbf{k}.\end{aligned}$$

Moments and Products of Inertia. Coördinates of a Self-conjugate Linear Vector=Function. The scalar coefficients which occur in the above expansion and are reprinted below, assuming for definiteness $A > B > C$,

$$\begin{aligned} A &\equiv \Sigma m(y^2+z^2), & D &\equiv \Sigma m\, yz, \\ B &\equiv \Sigma m(z^2+x^2), & E &\equiv \Sigma m\, zx, \qquad (227) \\ C &\equiv \Sigma m(x^2+y^2), & F &\equiv \Sigma m\, xy, \end{aligned}$$

are called the Moments of Inertia about the axes x, y, and z, and Products of Inertia with respect to the planes of yz, zx, and xy, respectively. The quantities A, B, and C are essentially positive, but D, E, and F may be of either sign. They may be found in any particular case by integration, for instance, since

$$dm = \rho\, dx\, dy\, dz,$$

and

$$\left.\begin{aligned} A &\equiv \iiint_{\text{Body}} (y^2+z^2)\, \rho\, dx\, dy\, dz \\ D &\equiv \iiint_{\text{Body}} yz\, \rho\, dx\, dy\, dz, \end{aligned}\right\} \qquad (228)$$

where ρ is the density at any point of the body.

Knowing these six coefficients (A, B, C, D, E, F), the function $\boldsymbol{\phi\omega}$ is completely determined, and for this reason they are sometimes called the coördinates of the self-conjugate linear vector-function.

Consider now the ellipsoid (224)

$$\boldsymbol{\omega}\cdot\boldsymbol{\phi\omega} = 2\,T = \text{const.}$$

$$\begin{aligned}\boldsymbol{\omega}\cdot\boldsymbol{\phi\omega} = \quad & \omega_1^2 A - \omega_1\,\omega_2 F - \omega_1\omega_3 E \\ & - \omega_2\omega_1 F + \omega_2^2\, B - \omega_2\omega_3 D \\ & - \omega_3\omega_1 E - \omega_3\,\omega_2 D + \omega_3^2\, C,\end{aligned} \tag{229}$$

which may also be written

$$\boldsymbol{\omega}\cdot\boldsymbol{\phi\omega} = \omega_1^2 A + \omega_2^2 B + \omega_3^2 C - 2\,\omega_2\omega_3 D - 2\,\omega_3\omega_1 E - 2\omega_1\omega_2 F.$$

It may be easily seen that in order that $\boldsymbol{\tau}\cdot\boldsymbol{\phi\sigma} = \boldsymbol{\sigma}\cdot\boldsymbol{\phi\tau}$, it is necessary that the coefficients D, E, F should occur in pairs as above. If however they do not occur in pairs, there will be nine coefficients in (229) all different. In this case the function is not said to be self-conjugate; but it is still a *linear* vector-function. In this last case $\boldsymbol{\tau}\cdot\boldsymbol{\phi\sigma}$ is not equal to $\boldsymbol{\sigma}\cdot\boldsymbol{\phi\tau}$. If we write $\boldsymbol{\tau}\cdot\boldsymbol{\phi\sigma} = \boldsymbol{\sigma}\cdot\boldsymbol{\phi}'\boldsymbol{\tau}$, $\boldsymbol{\phi}'$ is said to be the conjugate of $\boldsymbol{\phi}$, and $\boldsymbol{\phi}$ the conjugate of $\boldsymbol{\phi}'$.

Principal Moments of Inertia. Principal Axes. The function $\boldsymbol{\omega}\cdot\boldsymbol{\phi\omega}$, as is seen by its expansion, is homogeneous in $\boldsymbol{\omega}$, and the ellipsoid it represents when equated to a positive constant, $2\,T$, is referred to an origin at its center. We may now refer the ellipsoid to its three *principal* axes, its equation then becoming, as is well known,

$$\overline{A}\omega_1^2 + \overline{B}\omega_2^2 + \overline{C}\omega_3^2 = \text{const.} \tag{230}$$

The axes have lengths proportional to $\dfrac{1}{\sqrt{\overline{A}}}$, $\dfrac{1}{\sqrt{\overline{B}}}$, and $\dfrac{1}{\sqrt{\overline{C}}}$.

The coefficients $\overline{A}$, $\overline{B}$, and $\overline{C}$, in the above equation called the principal moments of inertia, are the moments of inertia of the body about these three principal axes and in general differ from the values they had before, but they are defined in the same manner with respect to the new axes. The products of inertia have all vanished. There are thus three directions in any rigid body for which the products of inertia when referred to them vanish.

Referred to these axes, since

$$\boldsymbol{\omega}\cdot\boldsymbol{\phi}\,\boldsymbol{\omega} \equiv \overline{A}\omega_1^2 + \overline{B}\omega_2^2 + \overline{C}\omega_3^2,$$

$\boldsymbol{\phi}\,\boldsymbol{\omega}$ then becomes $\quad \boldsymbol{\phi}\,\boldsymbol{\omega} \equiv \overline{A}\omega_1\mathbf{i} + \overline{B}\omega_2\mathbf{j} + \overline{C}\omega_3\mathbf{k}.$

and

$$\mathbf{H} = \overline{A}\omega_1\mathbf{i} + \overline{B}\omega_2\,\mathbf{j} + \overline{C}\omega_3\mathbf{k}. \tag{231}$$

So that the components of $\mathbf{H}$ are

$$H_x = \overline{A}\omega_1,$$
$$H_y = \overline{B}\omega_2,$$
$$H_z = \overline{C}\omega_3.$$

Looking upon $\boldsymbol{\phi}\,(\;)$ as an operator, we see from equation (231) that when it is applied to any vector $\boldsymbol{\omega}$

$$\boldsymbol{\omega} = \omega_1\mathbf{i} + \omega_2\mathbf{j} + \omega_3\mathbf{k},$$

thus $\quad \boldsymbol{\phi}\,\boldsymbol{\omega} = \overline{A}\omega_1\mathbf{i} + \overline{B}\omega_2\mathbf{j} + \overline{C}\omega_3\mathbf{k},$

it *multiplies* the components of $\boldsymbol{\omega}$ by the quantities $\overline{A}$, $\overline{B}$, and $\overline{C}$, respectively. Applying $\boldsymbol{\phi}$ again to $\boldsymbol{\phi}(\boldsymbol{\omega})$ we should obtain from

$$\boldsymbol{\phi\omega} = \overline{A}\omega_1\mathbf{i} + \overline{B}\omega_2\mathbf{j} + \overline{C}\omega_3\mathbf{k},$$
$$\boldsymbol{\phi\phi\omega} \equiv \boldsymbol{\phi}^2\boldsymbol{\omega} = \overline{A}^2\omega_1\mathbf{i} + \overline{B}^2\omega_2\mathbf{j} + \overline{C}^2\omega_3\mathbf{k}, \tag{232}$$

and so on. Defining $\boldsymbol{\phi}^{-1}$ as that operator which when applied to $\boldsymbol{\phi}$ annuls its effect, $\boldsymbol{\phi}^{-1}$ must then evidently *divide* the components of any vector by $\overline{A}$, $\overline{B}$, and $\overline{C}$, respectively, so that

$$\begin{aligned}\boldsymbol{\phi}^{-1}\boldsymbol{\phi\omega} = \boldsymbol{\omega} &= \boldsymbol{\phi}^{-1}\{\overline{A}\omega_1\mathbf{i} + \overline{B}\omega_2\mathbf{j} + \overline{C}\omega_3\mathbf{k}\}\\ &= \frac{1}{\overline{A}}\overline{A}\omega_1\mathbf{i} + \frac{1}{\overline{B}}\overline{B}\omega_2\mathbf{j} + \frac{1}{\overline{C}}\overline{C}\omega_3\mathbf{k}\\ &= \omega_1\mathbf{i} + \omega_2\mathbf{j} + \omega_3\mathbf{k},\end{aligned} \tag{233}$$

so that $\quad \boldsymbol{\phi}^{-1}\boldsymbol{\omega} = \dfrac{\omega_1}{\overline{A}}\mathbf{i} + \dfrac{\omega_2}{\overline{B}}\mathbf{j} + \dfrac{\omega_3}{\overline{C}}\mathbf{k}.$

Applying $\boldsymbol{\phi}^{-1}$ to this we obtain

$$\boldsymbol{\phi}^{-1}(\boldsymbol{\phi}^{-1}\boldsymbol{\omega}) \equiv \boldsymbol{\phi}^{-2}\omega = \frac{\omega_1}{A^2}\,\mathbf{i} + \frac{\omega_2}{B^2}\,\mathbf{j} + \frac{\omega_3}{C^2}\,\mathbf{k}.$$

and so on.

Lemma. We shall now show that $\boldsymbol{\phi}\boldsymbol{\omega}$ is perpendicular from the origin to the tangent plane at $\boldsymbol{\omega}$, and that its magnitude is inversely proportional to the distance from the origin to this tangent plane, or, in other words, that $(\boldsymbol{\phi}\boldsymbol{\omega})^{-1}$ * is the perpendicular vector from the origin to the tangent plane, at $\boldsymbol{\omega}$ of the quadric

$$\boldsymbol{\omega}\cdot\boldsymbol{\phi}\boldsymbol{\omega} = 1.$$

Consider the quadric

$$\boldsymbol{\omega}\cdot\boldsymbol{\phi}\boldsymbol{\omega} = \text{const.}$$

Differentiate this function, considering $\boldsymbol{\omega}$ as a variable

$$d\boldsymbol{\omega}\cdot\boldsymbol{\phi}\boldsymbol{\omega} + \boldsymbol{\omega}\cdot\boldsymbol{\phi}d\boldsymbol{\omega} = 2\,d\boldsymbol{\omega}\cdot\boldsymbol{\phi}\boldsymbol{\omega} = 0,$$

using c and d of equations (223).

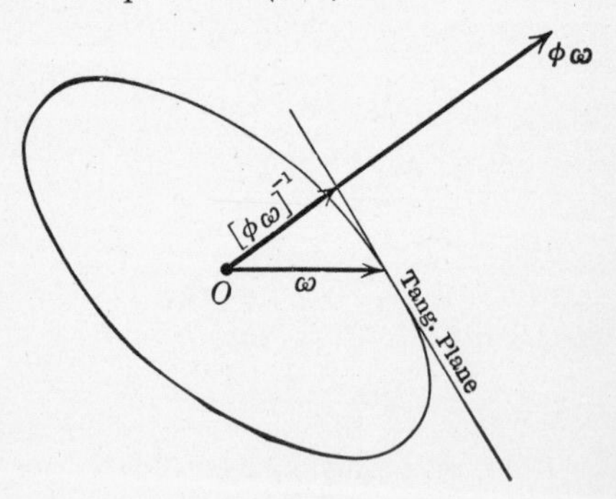

Fig. 65.

Hence $\boldsymbol{\phi}\boldsymbol{\omega}$ is perpendicular to $d\boldsymbol{\omega}$. But $d\boldsymbol{\omega}$ is a small vector in the surface at the extremity of $\boldsymbol{\omega}$ and therefore lies in the tangent plane, so that $\boldsymbol{\phi}\boldsymbol{\omega}$ is perpendicular to this plane. If $\boldsymbol{\sigma}$ is the running coördinate of the plane its equation is (64)

$$\boldsymbol{\sigma}\cdot\boldsymbol{\phi}\boldsymbol{\omega} = \text{const.},$$

* Do not confound $\boldsymbol{\phi}^{-1}(\ )$, the reciprocal operator, and $[\boldsymbol{\phi}(\ )]^{-1}$, the reciprocal of $\boldsymbol{\phi}$.

and by (68*a*) the perpendicular vector from the origin to this plane is

$$\left.\begin{array}{l}\mathbf{p} = \dfrac{\text{const.}}{\boldsymbol{\phi\omega}} = \text{const.}\,(\boldsymbol{\phi\omega})^{-1}. \\ \text{If the constant is unity,} \quad \mathbf{p} = (\boldsymbol{\phi\omega})^{-1}.\end{array}\right\} \qquad (234)$$

Physically this means, referring to (231), that the angular momentum $\mathbf{H} = \boldsymbol{\phi\omega}$ is normal to the tangent plane at $\boldsymbol{\omega}$ and inversely proportional to the length of the perpendicular from the origin on it.

Axes of a Central Quadric. The principal axes of a central quadric may be defined as those directions for which the magnitude of the radius vector is a maximum or a minimum. That is,

$$\boldsymbol{\omega}^2$$

is to be a max. or min. subject to the condition that

$$\boldsymbol{\omega}\cdot\boldsymbol{\phi\omega} = \text{const.}$$

Multiplying the first by an arbitrary multiplier and adding, the condition is obtained by writing the derivative of

$$\boldsymbol{\omega}\cdot(\boldsymbol{\phi\omega} - \lambda\boldsymbol{\omega}) = \text{const.}$$

to zero; this is

$$d\boldsymbol{\omega}\cdot(\boldsymbol{\phi\omega} - \lambda\boldsymbol{\omega}) + \boldsymbol{\omega}\cdot(\boldsymbol{\phi}d\boldsymbol{\omega} - \lambda d\boldsymbol{\omega}) = 0.$$

This becomes, using (223) (*c*), and (*d*),

$$d\boldsymbol{\omega}\cdot(\boldsymbol{\phi\omega} - \lambda\boldsymbol{\omega}) = 0$$

and since this must be true, independently as to how $\boldsymbol{\omega}$ varies, *i.e.*, it is to be a true maximum or minimum, it follows that

$$\boldsymbol{\phi\omega} = \lambda\boldsymbol{\omega}$$

is the condition required.

This is already an interesting result, for it states (234) that the radius-vector being parallel to $\boldsymbol{\phi\omega}$, is therefore *for those directions*, perpendicular to the surface. We might have started with this condition as a definition of principal axes.

This last equation is by (226) and (227) equivalent to

$$\begin{aligned} (A-\lambda)\omega_1 - \quad F\,\omega_2 - \quad E\,\omega_3 &= 0, \\ -F\,\omega_1 + (B-\lambda)\,\omega_2 - \quad D\,\omega_3 &= 0, \\ -E\,\omega_1 - \quad D\,\omega_2 + (C-\lambda)\,\omega_3 &= 0. \end{aligned} \tag{a}$$

The condition that these equations shall be compatible for values of ω_1, ω_2, ω_3, other than zero, is that the determinant of the coefficients shall vanish, *i.e.*,

$$\begin{vmatrix} A-\lambda & -F & -E \\ -F & B-\lambda & -D \\ -E & -D & C-\lambda \end{vmatrix} = 0. \tag{b}$$

This is a cubic in λ, and may be shown to always have three real roots. Each of these three values for λ inserted into equations (a) will allow for their solution, obtaining from each of them values for ω_1, ω_2 and ω_3 and hence a

$$\boldsymbol{\omega} = \omega_1\,\mathbf{i} + \omega_2\,\mathbf{j} + \omega_3\,\mathbf{k}$$

direction for each λ.

There are then always at least three principal axes to a central quadric surface.

The Principal Axes Intersect Normally. Let λ_1, λ_2, λ_3 be the roots of the determinantal cubic (b), and $\boldsymbol{\omega}_1$, $\boldsymbol{\omega}_2$, $\boldsymbol{\omega}_3$ the corresponding axes.

Then $$\boldsymbol{\phi}\boldsymbol{\omega}_1 = \lambda_1\boldsymbol{\omega}_1$$
and $$\boldsymbol{\phi}\boldsymbol{\omega}_2 = \lambda_2\boldsymbol{\omega}_2.$$

Multiplying the first by $\boldsymbol{\omega}_2\cdot$, and the second by $\boldsymbol{\omega}_1\cdot$ and substracting, there results, using (223) (*d*),

$$(\lambda_1 - \lambda_2)\ (\boldsymbol{\omega}_1\cdot\boldsymbol{\omega}_2) = 0,$$

which means that if λ_1 is not equal to λ_2, then the two principal axes $\boldsymbol{\omega}_1$ and $\boldsymbol{\omega}_2$ are perpendicular to each other; similarly if
$$\lambda_2 \neq \lambda_3 \text{ and } \lambda_3 \neq \lambda_1$$
we could show that the three principal axes are mutually perpendicular to each other.

If two roots of the cubic are equal, the position of the corresponding axes becomes indeterminate, and it may be shown that all radii perpendicular to the direction given by the third root are principal axes of the same length. The surface is then one of revolution about the determinate axis. If all three roots are equal, the surface is a sphere, and any axis is a principal axis. It may also be shown that the three roots of the cubic are equal to the squares of the reciprocals of the lengths of the semi-axes, the Cartesian equation then being

$$\lambda_1\omega_1^2 + \lambda_2\omega_2^2 + \lambda_3\omega_3^2 = \text{const.}$$

Comparing with (230) we see that the roots of the determinantal cubic are proportional to the principal moments of inertia,

$$\frac{\lambda_1}{A} = \frac{\lambda_2}{B} = \frac{\lambda_3}{C}.$$

89. Geometrical Representation of the Motion. Invariable Plane. If no impressed forces act upon the rotating body the equation of motion (218) becomes

$$\frac{d\mathbf{H}}{dt} = 0,$$

the solution of which is

$$\mathbf{H} = \boldsymbol{\phi}\boldsymbol{\omega} = \text{const. vector}, \qquad (235)$$

hence $\mathbf{H}$ or $\boldsymbol{\phi}\boldsymbol{\omega}$ is a vector constant in magnitude and direction throughout the motion, so that the tangent plane of the ellipsoid to which it is always perpendicular must remain fixed in space and is for this reason called the Invariable Plane. The point where this plane is touched by the ellipsoid is on the extremity of the instantaneous axis or pole, so that the ellipsoid is always rolling without sliding on this plane. In other words, having constructed the ellipsoid of inertia, and having determined the position of the invariable tangent plane in space, the motion of the body is the same as if it were rigidly attached to this ellipsoid which is rolling without sliding on the invariable plane.

This geometrical condition, in addition to the fact that the angular velocity of rotation is proportional to the radius vector to the point of contact of the ellipsoid and plane, completely determines the motion.

It is easy to see, since the *radius vector* always passes through a fixed point, the origin, that it must describe a cone in space, the vector **H** being its axis fixed in space. For this reason, **H**, is called the **Invariable Line.**

It also describes a cone in the ellipsoid. The vector **H** describes a cone in the ellipsoid because the ellipsoid moves relatively to it. This description of the motion is due to **Poinsot.**

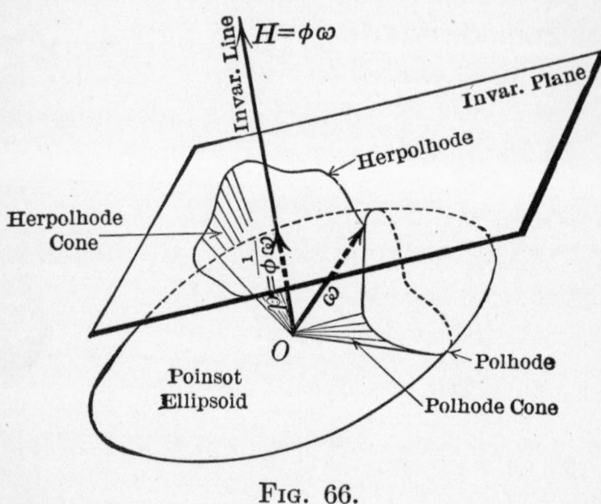

FIG. 66.

90. Polhode and Herpolhode Curves. If the paths described by the point of contact of the ellipsoid and invariable plane be determined on them, for instance by placing carbon paper between them as they roll on each other, two curves are obtained: one on the invariable plane, called the Herpolhode (sinuous path),* and one on the surface of the ellipsoid, called the Polhode (path of the pole). These curves are the directing curves of the cones described by the radius vector **ω** in space and in the ellipsoid, called respectively Herpolhode Cone and Polhode Cone.

Permanent Axes. It is easy to see that in three cases **H** and **ω** coincide in direction, *i.e.*, when **ω** is perpendicular to the tangent plane; in this case when both **H** and **ω** coincide in

* The above name is a misconception, because as a matter of fact the Herpolhode can be proved to have no point of inflection, and hence is not "sinuous."

direction along some one of the three principal axes of the momental ellipsoid these curves reduce to points and the ellipsoid rotates without rolling, permanently about these axes. There are then at least three directions at every point in a body about which if the body be set rotating it will continue to do so forever. Further consideration shows that two of these permanent axes are stable and one unstable, this last being the mean axis. The most stable is the least axis.

Equations of Polhode and Herpolhode Curves. The intersection of the cones described in the ellipsoid by the instantaneous axis with its surface will determine the polhode curves. In the quadric

$$\boldsymbol{\omega}\cdot\boldsymbol{\phi\omega} = \text{const.}$$

$\boldsymbol{\omega}$ must always satisfy the condition that the distance, p, to the tangent plane is constant, or that

$$\left(\frac{1}{\boldsymbol{\phi\omega}}\right)^2 = \mathbf{p}^2 = \text{const.},$$

or $$(\boldsymbol{\phi\omega})^2 = \frac{1}{\mathbf{p}^2} = \boldsymbol{\phi\omega}\cdot\boldsymbol{\phi\omega} = \boldsymbol{\omega}\cdot\boldsymbol{\phi}(\boldsymbol{\phi\omega}) = \boldsymbol{\omega}\cdot\boldsymbol{\phi}^2\boldsymbol{\omega} \qquad \text{(a quadric)},$$

where we define $\boldsymbol{\phi}^2(\ ) = \boldsymbol{\phi}(\boldsymbol{\phi}(\ ))$ etc., see (232),

so that the two equations

$$\left.\begin{aligned} \boldsymbol{\omega}\cdot\boldsymbol{\phi\omega} &= \text{const.} = k \\ \text{and} \qquad \boldsymbol{\omega}\cdot\boldsymbol{\phi}^2\boldsymbol{\omega} &= \frac{1}{\mathbf{p}^2} \end{aligned}\right\} \tag{236}$$

must be simultaneously satisfied. Combining them, we obtain from

$$\frac{1}{\mathbf{p}^2}\,\boldsymbol{\omega}\cdot\boldsymbol{\phi\omega} = \frac{k}{\mathbf{p}^2} \quad \text{and} \quad k\,\boldsymbol{\omega}\cdot\boldsymbol{\phi}^2\boldsymbol{\omega} = \frac{k}{\mathbf{p}^2}$$

by subtraction

$$k\boldsymbol{\omega}\cdot\boldsymbol{\phi}^2\boldsymbol{\omega} - \frac{1}{\mathbf{p}^2}\,\boldsymbol{\omega}\cdot\boldsymbol{\phi\omega} = 0,$$

or finally

$$\boldsymbol{\omega}\cdot\boldsymbol{\phi}\left(k\boldsymbol{\phi} - \frac{1}{\mathbf{p}^2}\right)\boldsymbol{\omega} = 0, \tag{237}$$

a homogeneous equation of the second degree, and hence a cone with vertex at the origin. Its equation in Cartesian coördinates may be immediately written down by (231) and (232) as

$$\omega_1^2\left(\bar{A}^2 k - \frac{\bar{A}}{p^2}\right) + \omega_2^2\left(\bar{B}^2 k - \frac{\bar{B}}{p^2}\right) + \omega_3^2\left(\bar{C}^2 k - \frac{\bar{C}}{p^2}\right) = 0,$$

or (238)

$$\bar{A}\left(\bar{A}k - \frac{1}{p^2}\right)\omega_1^2 + \bar{B}\left(\bar{B}k - \frac{1}{p^2}\right)\omega_2^2 + \bar{C}\left(\bar{C}k - \frac{1}{p^2}\right)\omega_3^2 = 0.$$

The intersections of this cone for *different* values of $\mathbf{p}$ with the ellipsoid

$$\boldsymbol{\omega}\cdot\boldsymbol{\phi}\boldsymbol{\omega} = k,$$

give the polhode curves, which are, therefore, twisted curves of the fourth degree, lying on the momental or Poinsot ellipsoid.

Since the herpolhode is traced out by the points of contact of an ellipsoid rotating on its center with an invariable tangent plane, these curves must lie between two concentric circles on the plane, their centers being at the intersection of the invariable line $\mathbf{H}$ with that plane, and touching them alternately.

Moving Axes and Relative Motion.

91. Theorem of Coriolis. It is often convenient in dynamics to use axes which themselves move in space and to which the motions of the body under consideration are referred.

In order to determine at any time the position of the moving axes, one method is to refer them to axes which remain at rest throughout the motion. According to this device the fixed axes are left behind by the moving ones. However, it is found to be more advantageous to refer the moving axes at all times to fixed axes instantaneously coinciding with them.

No generality is lost by referring the motion of a body to moving axes which simply turn about a *fixed* point in space, as any motion of translation of the moving axes with reference to fixed ones may be compensated for by giving to every point of the body considered a motion equal and opposite to that of these moving axes. This condition, then, does not limit the generality of the choice of moving axes.

Consider any vector $\overrightarrow{OP} = \mathbf{r}$ drawn from a fixed origin O and for definiteness let it be the vector to a point P in a moving body. We shall now consider the motion of the

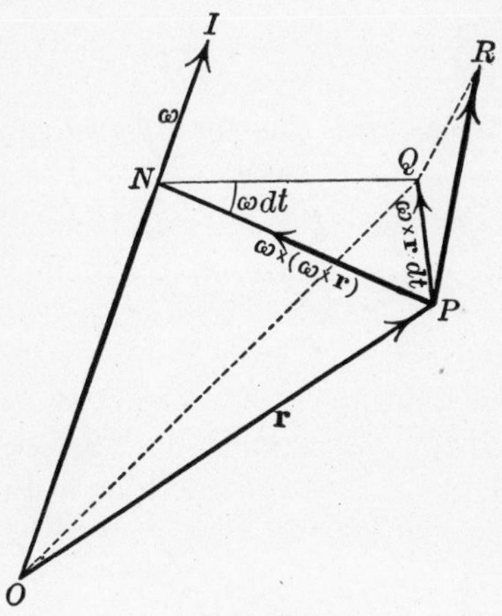

FIG. 67.

point P in two ways. Refer it to two different *spaces* initially coincident, one revolving about the axis OI with an angular velocity $\boldsymbol{\omega}$, the other remaining at rest (or fixed).

Let PR be the motion seen in a time dt by an observer remaining in the fixed space, or, in other words, the absolute motion in space. Denote this vector $\overrightarrow{PR}$ as $d\mathbf{r}_{fs}$, the subscript denoting its reference to fixed space. Consider now the point P as remaining at rest with reference to the moving space; it will therefore move relatively to fixed space with the velocity of the moving space alone and will describe the path

$$PQ = \boldsymbol{\omega} dt \times \mathbf{r} = \boldsymbol{\omega} \times \mathbf{r}\, dt,$$

as $\boldsymbol{\omega} dt$ is the angle described by NP in time dt.

But as the particle P by its own motion actually reaches the point R, the vector $\overline{QR}$ must represent the path of the particle as seen by an observer moving with the moving space; in other words, $\overline{QR} = d\mathbf{r}_{ms}$, the subscript denoting its reference to moving space. From the figure

$$d\mathbf{r}_{fs} = d\mathbf{r}_{ms} + \boldsymbol{\omega}\times\mathbf{r}\,dt.$$

Dividing through by dt we obtain the very important equation

$$\left(\frac{d\mathbf{r}}{dt}\right)_{fs} = \left(\frac{d\mathbf{r}}{dt}\right)_{ms} + \boldsymbol{\omega}\times\mathbf{r}. \qquad (239)$$

Letting $OP = \mathbf{r}$ represent any directed quantity such as force, velocity, moment of a couple, or angular momentum, etc., equation (239) shows how to refer them to a moving space.

The vector $\mathbf{r}$ always represents the vector at the beginning of the motion and referred to either space, as initially they are both coincident. If $\mathbf{r}$ represent a displacement and $\mathbf{q}$ the velocity of the point P,

$$\mathbf{q}_{fs} = \left(\frac{d\mathbf{r}}{dt}\right)_{ms} + \boldsymbol{\omega}\times\mathbf{r}.$$

The acceleration of a body whose motion is known relatively to moving space, and the motion of moving space known relatively to fixed space, may be obtained by a second application of this equation.

Replace $\mathbf{r}$ by $\mathbf{q}_{fs}$ thus,

$$\mathbf{a}_{fs} = \left(\frac{d\mathbf{q}_{fs}}{dt}\right)_{ms} + \boldsymbol{\omega}\times\mathbf{q}_{fs}$$

$$= \left(\frac{d^2\mathbf{r}}{dt^2}\right)_{ms} + \frac{d}{dt}(\boldsymbol{\omega}\times\mathbf{r}) + \boldsymbol{\omega}\times\left(\frac{d\mathbf{r}}{dt}\right)_{ms} + \boldsymbol{\omega}\times(\boldsymbol{\omega}\times\mathbf{r})$$

$$= \left(\frac{d^2\mathbf{r}}{dt^2}\right)_{ms} + 2\,\boldsymbol{\omega}\times\left(\frac{d\mathbf{r}}{dt}\right)_{ms} + \frac{d\boldsymbol{\omega}}{dt}\times\mathbf{r} + \boldsymbol{\omega}\times(\boldsymbol{\omega}\times\mathbf{r}). \qquad (240)$$

This equation will readily expand to the familiar ones referred to Cartesian axes, as ordinarily given, for example:

$$(a_{fs})_x = \left(\frac{d^2x}{dt^2}\right)_{ms} + 2\left(\omega_2\frac{dz}{dt} - \omega_3\frac{dy}{dt}\right) + \left(z\frac{d\omega_2}{dt} - y\frac{d\omega_3}{dt}\right) - x\,(\omega_2^{\,2} + \omega_3^{\,2})$$
$$+\omega_1\omega_2 z + \omega_1\omega_2 y. \qquad (241)$$

The last three terms may be written out as above by remembering that

$$\boldsymbol{\omega}\times(\boldsymbol{\omega}\times\mathbf{r}) = \boldsymbol{\omega}(\boldsymbol{\omega}\cdot\mathbf{r}) - \mathbf{r}\,\boldsymbol{\omega}^2.$$

If the point P is *attached* to the moving space,

$$\left(\frac{d^2\mathbf{r}}{dt^2}\right)_{ms} = 0, \text{ and also } 2\,\boldsymbol{\omega}\times\left(\frac{d\mathbf{r}}{dt}\right)_{ms} = 0,$$

so that the remaining expressions are the accelerations produced by the motion of the moving space itself.

Hence the so-called

$$\textit{Acceleration of moving space} = \frac{d\boldsymbol{\omega}}{dt}\times\mathbf{r} + \boldsymbol{\omega}\times(\boldsymbol{\omega}\times\mathbf{r}). \qquad (242)$$

If the angular velocity of moving space is constant,

$$\frac{d\boldsymbol{\omega}}{dt} = 0 \text{ and therefore } \frac{d\boldsymbol{\omega}}{dt}\times\mathbf{r} = 0.$$

The remaining term $\boldsymbol{\omega}\times(\boldsymbol{\omega}\times\mathbf{r})$ is the acceleration produced on the body by its individual rotation about the axis OI. Since $\boldsymbol{\omega}\times\mathbf{r}$ is perpendicular to $\boldsymbol{\omega}$, $\boldsymbol{\omega}\times(\boldsymbol{\omega}\times\mathbf{r})$ is perpendicular both to $\boldsymbol{\omega}\times\mathbf{r}$ and to $\boldsymbol{\omega}$ and directed normally towards the axis $\boldsymbol{\omega}$. This is the ordinary Centripetal Acceleration. Because in the vector product $\boldsymbol{\omega}\times\mathbf{r}$, $\overline{NP}$ instead of $\mathbf{r}$ may be used (see § 36) without changing its value, and as $\boldsymbol{\omega}$ and $\overline{NP}$ are at right angles,

$$\boldsymbol{\omega}\cdot\overline{NP} = 0,$$

this centripetal acceleration may then be written

$$\boldsymbol{\omega}\times(\boldsymbol{\omega}\times\mathbf{r}) = -\boldsymbol{\omega}^2\overline{NP}.$$

The term $2\,\boldsymbol{\omega}\times\left(\frac{d\mathbf{r}}{dt}\right)_{ms}$ is called the *Compound Centripetal Acceleration of Coriolis.*

We may then consider the moving axes to be at rest if to the *actual* forces applied to the body fictitious ones be added capable of producing accelerations equal and opposite to the acceleration of moving space and to the compound centripetal acceleration. This is the theorem of **Coriolis.**

92. Transformation of the Equation of Motion. Centrifugal Couple. Let us utilize equation (239) to refer the motion of a rigid body to a space moving with it, or, in other words, to axes in the body. The equation of motion of a rigid body about a point referred to fixed space is (218)

$$\frac{d\mathbf{H}}{dt} = \mathbf{M}.$$

If there are no impressed forces, $\mathbf{M} = 0$ and

$$\frac{d\mathbf{H}}{dt} = 0$$

states that $\mathbf{H}$, the moment of momentum, remains constant in magnitude and direction in the fixed space, however peculiar the motion of the body may seem to be.

Employing now the equation of Coriolis, we substitute for $\left(\frac{d\mathbf{H}}{dt}\right)_{fs}$ its equivalent for moving space, obtaining as the equation of motion of a rigid body about a fixed point referred to a space moving with the body

$$\left(\frac{d\mathbf{H}}{dt}\right)_{ms} + \boldsymbol{\omega}\times\mathbf{H} = \mathbf{M}. \tag{243}$$

If there are no applied forces, $\mathbf{M} = 0$, and (243) becomes

$$\left(\frac{d\mathbf{H}}{dt}\right)_{ms} = \mathbf{H}\times\boldsymbol{\omega}. \tag{244}$$

$\mathbf{H}\times\boldsymbol{\omega}$ is called the Centrifugal Couple and, as is evident by its form, is perpendicular to both $\boldsymbol{\omega}$ and $\mathbf{H}$. The above equation then states that the rate of change of the angular momentum $\mathbf{H}$ *in the body* is equal to the centrifugal couple, $\mathbf{H}\times\boldsymbol{\omega}$. If the change in $\mathbf{H}$ is always normal to itself, then $\mathbf{H}$ is never increased or decreased in length but only changes in direction at a rate proportional to $\mathbf{H}\times\boldsymbol{\omega}$. Thus $\mathbf{H}$ describes a cone in the body, although it remains fixed in fixed space.

If $\mathbf{H}$ and $\boldsymbol{\omega}$ ever become parallel then $\mathbf{H}\times\boldsymbol{\omega}$ vanishes, and *in the body* too we have

$$\left(\frac{d\mathbf{H}}{dt}\right)_{ms} = 0,$$

or, in other words, the body must continue to rotate forever about the Invariable Line $\mathbf{H}$ in fixed space; it is then an invariable line in the body also. We have seen that there are at least three such directions, called permanent axes, or principal axes, for which the above condition is fulfilled. A symmetrical body supported at its center of mass and rotating about its axis of symmetry will give this kind of motion.

Gyroscope. The property that a rotating body possesses of rotating permanently about a principal axis was utilized by Foucault in the gyroscope. When a symmetrical top is rapidly spinning in gimbals, it keeps its axis pointing in the same direction (invariable line) in space, so that if the top is carried around by the earth's motion, the axis of the top remaining fixed in fixed space will describe a cone with reference to the earth (or moving space). By observations on such an instrument not only can the rotation of the earth be proved but the latitude of the locality at which the experiment is performed may be determined.

93. Euler's Equations. If ω_1, ω_2, ω_3 be the three components of $\boldsymbol{\omega}$ along the principal axes of a rigid body at a point, and if $\overline{A}$, $\overline{B}$, $\overline{C}$ be the principal moments of inertia about those same axes, we may write (231)

$$\begin{aligned}\mathbf{H} &= \overline{A}\omega_1\mathbf{i} + \overline{B}\omega_2\mathbf{j} + \overline{C}\omega_3\mathbf{k}\\ &= \boldsymbol{\phi}\boldsymbol{\omega}.\end{aligned}$$

There are no products of inertia entering into this equation, as for the principal axes they vanish (§ 89).

Substituting this value for **H** in (243) there result the three equations

$$\left.\begin{aligned} \overline{A}\frac{d\omega_1}{dt} + (\overline{C} - \overline{B})\omega_2\omega_3 = M_1, \\ \overline{B}\frac{d\omega_2}{dt} + (\overline{A} - \overline{C})\omega_3\omega_1 = M_2, \\ \overline{C}\frac{d\omega_3}{dt} + (\overline{B} - \overline{A})\omega_1\omega_2 = M_3. \end{aligned}\right\} \qquad (245)$$

These are the dynamical equations of Euler for the motion of a rigid body about a fixed point, referred to axes moving with the body. Of course

$$\frac{d\mathbf{H}}{dt} + \boldsymbol{\omega}\times\mathbf{H} = \mathbf{M}$$

is the corresponding vector equation.

94. Analytical Solution of Euler's Equations for Motion under No Impressed Forces. For convenience we rewrite the following:

$$\boldsymbol{\omega} = \omega_1\mathbf{i} + \omega_2\mathbf{j} + \omega_3\mathbf{k}. \qquad (a)$$

$$\mathbf{H} = \overline{A}\omega_1\mathbf{i} + \overline{B}\omega_2\mathbf{j} + \overline{C}\omega_3\mathbf{k} = \boldsymbol{\phi}\boldsymbol{\omega}. \qquad (b)$$

Then $$\boldsymbol{\phi}^{-1}\boldsymbol{\omega} = \frac{\omega_1}{\overline{A}}\mathbf{i} + \frac{\omega_2}{\overline{B}}\mathbf{j} + \frac{\omega_3}{\overline{C}}\mathbf{k}. \qquad (c)$$

Also $$\frac{d}{dt}\boldsymbol{\phi}(\ ) = \boldsymbol{\phi}\frac{d}{dt}(\ ). \qquad (d)$$

$$\boldsymbol{\sigma}\cdot\boldsymbol{\phi}\boldsymbol{\omega} = \boldsymbol{\omega}\cdot\boldsymbol{\phi}\boldsymbol{\sigma}. \qquad (e)$$

Euler's equation for this case is

$$\frac{d\mathbf{H}}{dt} = \mathbf{H}\times\boldsymbol{\omega} \quad \text{or} \quad \frac{d\boldsymbol{\phi}\boldsymbol{\omega}}{dt} = \boldsymbol{\phi}\boldsymbol{\omega}\times\boldsymbol{\omega}. \qquad (246)$$

Since the right-hand side is perpendicular to $\boldsymbol{\phi\omega}$ and to $\boldsymbol{\omega}$, let us take the scalar products of the equation with $\boldsymbol{\phi\omega}$ and with $\boldsymbol{\omega}$ respectively. Multiplying with $\boldsymbol{\omega}\cdot$ we obtain

$$\boldsymbol{\omega}\cdot\frac{d\boldsymbol{\phi\omega}}{dt} = \boldsymbol{\omega}\cdot\boldsymbol{\phi\omega}\times\boldsymbol{\omega} = 0.$$

which we must integrate.
By differentiating $\boldsymbol{\omega}\cdot\boldsymbol{\phi\omega} =$ const. with respect to the time thus,

$$\boldsymbol{\omega}\cdot\frac{d}{dt}\boldsymbol{\phi\omega} + \frac{d\boldsymbol{\omega}}{dt}\cdot\boldsymbol{\phi\omega} = 0,$$

which becomes by (*e*) and (*d*)

$$\boldsymbol{\omega}\cdot\frac{d}{dt}\boldsymbol{\phi\omega} + \boldsymbol{\omega}\cdot\frac{d}{dt}\boldsymbol{\phi\omega} = 2\,\boldsymbol{\omega}\cdot\frac{d}{dt}\boldsymbol{\phi\omega} = 0$$

we see that the integral of the equation is

$$\boldsymbol{\omega}\cdot\boldsymbol{\phi\omega} = 2\,T, \tag{247}$$

where T is an arbitrary constant. Those having read the preceding pages will recognize in (247) the equation of the Poinsot Ellipsoid, and in T the kinetic energy of the body.

Multiplying the original equation with $\boldsymbol{\phi\omega}\cdot$ we have immediately

$$\boldsymbol{\phi\omega}\cdot\frac{d\boldsymbol{\phi\omega}}{dt} = \boldsymbol{\phi\omega}\cdot\boldsymbol{\phi\omega}\times\boldsymbol{\omega} = 0,$$

and by (*e*) and (*d*)

$$\tfrac{1}{2}\,d(\boldsymbol{\phi\omega})^2 = 0,$$

whose integral is

$$(\boldsymbol{\phi\omega})^2 = \mathbf{H}^2, \tag{248}$$

where $\mathbf{H}^2$ is an arbitrary constant. We recognize here the constancy of $\mathbf{H}$ in *magnitude* only in the body, hence any change in $\mathbf{H}$ must be perpendicular to it. See § 92.

In order to obtain a third integral, multiply (246) by $\boldsymbol{\phi}^{-1}\boldsymbol{\omega}$, the $\boldsymbol{\phi}^{-1}$ and the $\boldsymbol{\phi}$ annulling each other.

$$\boldsymbol{\phi}^{-1}\boldsymbol{\omega}\cdot\frac{d\boldsymbol{\phi\omega}}{dt} = \boldsymbol{\omega}\cdot\frac{d\boldsymbol{\omega}}{dt} = \boldsymbol{\phi}^{-1}\boldsymbol{\omega}\cdot\boldsymbol{\phi\omega}\times\boldsymbol{\omega}. \tag{249}$$

From these three equations (247), (248), (249) it is possible to find $\boldsymbol{\omega}$ for all time and hence all about the motion. Suppose it is desired to expand these three integrals into Cartesian form; we have immediately for (247) and (248)

$$\boldsymbol{\omega}\cdot\boldsymbol{\phi}\boldsymbol{\omega} = 2\,T = \bar{A}\omega_1^2 + \bar{B}\omega_2^2 + \bar{C}\omega_3^2 \tag{250}$$

and

$$(\boldsymbol{\phi}\boldsymbol{\omega})^2 = \mathbf{H}^2 = \bar{A}^2\omega_1^2 + \bar{B}^2\omega_2^2 + \bar{C}^2\omega_3^2. \tag{251}$$

The third is more complicated, but easy,

$$\boldsymbol{\omega}\cdot\frac{d\boldsymbol{\omega}}{dt} = \boldsymbol{\phi}^{-1}\boldsymbol{\omega}\cdot\boldsymbol{\phi}\boldsymbol{\omega}\times\boldsymbol{\omega}$$

$$= \left(\frac{\omega_1}{\bar{A}}\mathbf{i} + \frac{\omega_2}{\bar{B}}\mathbf{j} + \frac{\omega_3}{\bar{C}}\mathbf{k}\right)\cdot\begin{vmatrix} \mathbf{i} & \mathbf{j} & \mathbf{k} \\ \bar{A}\omega_1 & \bar{B}\omega_2 & \bar{C}\omega_3 \\ \omega_1 & \omega_2 & \omega_3 \end{vmatrix}$$

$$= \left\{\frac{\bar{B}-\bar{C}}{\bar{A}} + \frac{\bar{C}-\bar{A}}{\bar{B}} + \frac{\bar{A}-\bar{B}}{\bar{C}}\right\}\omega_1\omega_2\omega_3. \tag{252}$$

By solving equations (250) and (251) combined with

$$\omega^2 = \omega_1^2 + \omega_2^2 + \omega_3^2,$$

for ω_1, ω_2 and ω_3, and substituting in (252), we find

$$\omega\frac{d\omega}{dt} = \sqrt{(\lambda_1 - \omega^2)\,(\lambda_2 - \omega^2)\,(\lambda_3 - \omega^2)},$$

where the λ's are functions of $\bar{A}$, $\bar{B}$, $\bar{C}$, T and H, an equation for ω in terms of these constants whose general solution involves elliptic functions.*

95. Hamilton's Principle. Starting again with d'Alembert's equation,

$$\sum\left(m\frac{d^2\mathbf{r}}{dt^2} - \mathbf{F}\right)\cdot\delta\mathbf{r} = 0, \tag{253}$$

the $\delta\mathbf{r}$'s being any variations consistent with the constraints imposed upon the system, or, what is the same thing, satisfy

* See article by Professor Greenhill, in fourteenth volume of the *Quarterly Journal*, pp. 182 and 265, 1876.

certain equations of condition, we may transform it in the following manner:

$$\frac{d^2\mathbf{r}}{dt^2}\cdot\delta\mathbf{r} = \frac{d}{dt}\left(\frac{d\mathbf{r}}{dt}\cdot\delta\mathbf{r}\right) - \frac{d\mathbf{r}}{dt}\cdot\frac{d\delta\mathbf{r}}{dt}$$

$$= \frac{d}{dt}\left(\frac{d\mathbf{r}}{dt}\cdot\delta\mathbf{r}\right) - \delta\frac{1}{2}\left(\frac{d\mathbf{r}}{dt}\right)^2.$$

Treat each term of (253) in this way; d'Alembert's equation then becomes

$$\frac{d}{dt}\sum m\left(\frac{d\mathbf{r}}{dt}\cdot\delta\mathbf{r}\right) = \delta\frac{1}{2}\sum m\left(\frac{d\mathbf{r}}{dt}\right)^2 + \sum \mathbf{F}\cdot\delta\mathbf{r}$$

$$= \delta T + \sum \mathbf{F}\cdot\delta\mathbf{r},$$

where T is the kinetic energy due to the velocities of the masses of the system. As the first term is an exact derivative, let us integrate with respect to the time from $t = t_1$ to $t = t_2$,

$$\sum m\left(\frac{d\mathbf{r}}{dt}\cdot\delta\mathbf{r}\right)\Big/_{t_1}^{t_2} = \int_{t_1}^{t_2}\left(\delta T + \sum \mathbf{F}\cdot\delta\mathbf{r}\right)dt.$$

If the positions of the system are given at the times t_1 and t_2, then the $\delta\mathbf{r}$'s are zero for those times, and the left-hand terms vanish, leaving

$$\int_{t_1}^{t_2}\left(\delta T + \sum \mathbf{F}\cdot\delta\mathbf{r}\right)dt = 0. \qquad (254)$$

This equation is true whatever the system of forces is that acts; if, however, the system is a conservative one, the work done (by definition) in going from any point to any other point against these forces is independent of the path chosen, and is therefore a function solely of the initial and final points of the path. In this case, then, there must be a scalar point-function, ($-W$ say), such that knowing its value everywhere we can calculate the work done in going from any point to any other by any path, simply by knowing the values of $-W$, for those points. (Consult § 59.) In other words, the work

$\Sigma_1^2 \mathbf{F} \cdot \delta \mathbf{r}$ in going from position 1 to position 2 is the difference in value of the function $-W$ at positions 1 and 2, or

$$\Sigma_1^2 \mathbf{F} \cdot \delta \mathbf{r} = -(W_1 - W_2) = -\delta W.$$

In the case of conservative forces, then, $\Sigma\, \mathbf{F} \cdot \delta \mathbf{r}$ may be replaced by $-\delta W$ and (254) becomes

$$\delta \int_{t_1}^{t_2} (T - W)\, dt = 0. \tag{255}$$

This is **Hamilton's Principle.**

Lagrangian Function. The function $T - W$ is called the Lagrangian Function, and is often written L, so that Hamilton's integral becomes

$$\delta \int_{t_1}^{t_2} L\, dt = 0. \tag{256}$$

96. Extension of the Conception of Vector to More than Three Dimensions. Certain processes occur in mathematical physics in which more than three independent variables are concerned. In such cases as this the vector notation is still applicable to the manipulation of these quantities. If $q_1, q_2, q_3 \ldots$ be these independent quantities, we conceive of a vector $\mathbf{q}$,

$$\mathbf{q} = q_1 \mathbf{i}_1 + q_2 \mathbf{i}_2 + q_3 \mathbf{i}_3 + \cdots,$$

where $\mathbf{i}_1, \mathbf{i}_2, \mathbf{i}_3 \ldots$ are independent unit vectors. By an extension of the idea of a vector, we are to consider $\mathbf{q}$ as a vector existing in more than three dimensions, as many as there are q's. An example will show that we are not going very far beyond the manipulation of ordinary vectors.

Definitions. Consider the generalized vector in n-dimensions,

$$\mathbf{q} = q_1 \mathbf{i}_1 + q_2 \mathbf{i}_2 + \cdots + q_n \mathbf{i}_n. \tag{257}$$

In analogy to $\nabla \equiv \frac{\partial}{\partial x} \mathbf{i} + \frac{\partial}{\partial y} \mathbf{j} + \frac{\partial}{\partial z} \mathbf{k}$

write

$$\nabla_n \equiv \frac{\partial}{\partial q_1} \mathbf{i}_1 + \frac{\partial}{\partial q_2} \mathbf{i}_2 + \cdots + \frac{\partial}{\partial q_n} \mathbf{i}_n. \tag{258}$$

If we write $\mathbf{q}_s'$ for $\frac{\partial \mathbf{q}_s}{\partial t}$, we may have also

$$\mathbf{q}' = q_1'\mathbf{i}_1 + q_2'\mathbf{i}_2 + \cdots + q_n'\mathbf{i}_n \tag{259}$$

and define also ∇_n'

$$\nabla_n' \equiv \frac{\partial}{\partial q_1'}\mathbf{i}_1 + \frac{\partial}{\partial q_2'}\mathbf{i}_2 + \cdots + \frac{\partial}{\partial q_n'}\mathbf{i}_n. \tag{260}$$

Let us apply this notation to the following transformation.

97. Lagrange's General Equations of Motion. By means of Hamilton's Integral we may deduce Lagrange's Equations. Let the position $\mathbf{r}_s$ of any point s of the dynamical (or analogous) system be expressible in terms of the *independent* parameters, $q_1, q_2, \ldots q_n$, so that

$$\mathbf{r}_s = \boldsymbol{\phi}_s(q_1\, q_2\, q_3 \ldots q_n). \tag{261}$$

This means that to every point

$$\mathbf{q} = q_1\mathbf{i}_1 + q_2\mathbf{i}_2 + \cdots + q_n\mathbf{i}_n$$

in a space of n-dimensions there corresponds a definite value for *all* of the parameters of the dynamical system and hence a definite and determinable configuration of every particle in the system. If it is possible to find what functions of the time the q_s's are, subject to the dynamical equations of condition, included in Hamilton's principle, the position of every particle for all instants will be known. Hence the simple motion of *one point* in n-dimensions includes the problem of the motion of a *system* of points in three or less dimensions (or for that matter in more than three).

Differentiating (261),

$$\boldsymbol{\phi}_s' = \frac{d\mathbf{r}_s}{dt} = \frac{\partial \boldsymbol{\phi}_s}{\partial q_1} q_1' + \frac{\partial \boldsymbol{\phi}_s}{\partial q_2} q_2' + \cdots = \mathbf{q}' \cdot \nabla_n \boldsymbol{\phi}_s. \tag{262}$$

Thus every $\boldsymbol{\phi}'$ is expressible as a *linear* function of the q''s.

The kinetic energy function T becomes

$$T = \frac{1}{2} \sum m \left(\frac{d\mathbf{r}}{dt}\right)^2 = \frac{1}{2} \sum m \, (\mathbf{q}' \cdot \nabla_n \boldsymbol{\phi})^2, \tag{263}$$

every term of the sum being linear and homogeneous in q'. The square is a homogeneous quadratic function of the q''s; often written

$$T = \tfrac{1}{2} Q_{11} q'^2_1 + \tfrac{1}{2} Q_{22} q'^2_2 + \cdots + Q_{12} q_1' q_2' + \cdots \tag{264}$$

Performing the variation with respect to q' and q indicated by Hamilton's integral (255), and using $L \equiv T - W$, we have since

$$\delta L = \delta \mathbf{q} \cdot \nabla_n L + \delta \mathbf{q}' \cdot \nabla_n' L,$$

$$\int_{t_1}^{t_2} (\delta \mathbf{q} \cdot \nabla_n L + \delta \mathbf{q}' \cdot \nabla_n' L) \, dt = 0.$$

But because

$$\delta \mathbf{q}' = \delta \frac{d\mathbf{q}}{dt} = \frac{d}{dt} \delta \mathbf{q},$$

the second term becomes

$$\int_{t_1}^{t_2} \left[\left(\frac{d}{dt} \delta \mathbf{q} \right) \cdot \nabla_n' L \right] dt,$$

which may be integrated by parts, with respect to the time, into

$$\delta \mathbf{q} \cdot \nabla_n' L \Big|_{t_1}^{t_2} - \int_{t_1}^{t_2} \delta \mathbf{q} \cdot \frac{d}{dt} \nabla_n' L,$$

of which the first term vanishes, as $\mathbf{q}$ is fixed at the limits and hence suffers no variation there, leaving altogether

$$\int_{t_1}^{t_2} \left\{ \delta \mathbf{q} \cdot \left(\nabla_n L - \frac{d}{dt} \nabla_n' L \right) \right\} dt = 0.$$

But as $\delta \mathbf{q}$ is arbitrary, it follows that

$$\frac{d}{dt} \nabla_n' L - \nabla_n L = 0. \tag{265}$$

This is a vector in n-dimensional space, whose components must all be zero, for example:

$$\left.\begin{aligned} \frac{d}{dt}\left(\frac{\partial(T-W)}{\partial q_1'}\right)-\frac{\partial(T-W)}{\partial q_1}=0, \\ \frac{d}{dt}\left(\frac{\partial(T-W)}{\partial q_2'}\right)-\frac{\partial(T-W)}{\partial q_2}=0, \text{ etc.} \end{aligned}\right\} \qquad (266)$$

These are **Lagrange's Generalized Equations of Motion.**

If then for any system the functions T and W are known, and it is possible to express them in terms of n-independent parameters $q_1, q_2, \ldots q_n$, these n-equations (266) make it possible to determine the values of these parameters for all time; thus the path of the point

$$\mathbf{q} = q_1\mathbf{i}_1 + q_2\mathbf{i}_2 + \cdots q_n\mathbf{i}_n$$

is determined, in n-dimensional space.

Defining the operator $\overline{\nabla}(\ )$,

$$\overline{\nabla}(\ ) \equiv \left(\frac{d}{dt}\nabla_n' - \nabla_n\right)(\ ),$$

all dynamics is included under the **Remarkable Formula**

$$\overline{\nabla} L = 0. \qquad (267)$$

Hydrodynamics.

98. Fundamental Equations. We shall now derive the fundamental equations of hydrodynamics for a frictionless fluid and some of their most important consequences by means of the previous principles. The directness of attack and absence of artificiality is especially noticeable in this application of the vector method.

Let ρ represent the density of the fluid; we shall assume that it is a function of the pressure p alone, so that

$$\rho = f(p). \qquad (268)$$

Equation of Continuity. Let $\mathbf{q}$ $(u\,v\,w)$ be the velocity of the fluid and $\mathbf{F}(X\,Y\,Z)$ be the force per unit mass acting on the fluid. Consider a *fixed* surface S in the fluid. By fixed is meant that the imagined surface retains its position in space irrespective of the motion of the fluid itself. The rate of increase of matter in it is measured by the surface integral of the flux of the fluid though the surface taken along the *inward* drawn normal. No fluid is supposed to be created nor destroyed inside of the surface. As we use generally the outward drawn normal $\mathbf{n}$ in our formulæ, $-\mathbf{n}$ will represent the inward drawn normal, so that we may then write using the divergence theorem

$$\frac{\partial m}{\partial t} = -\iint_S \rho\, \mathbf{q}\cdot\mathbf{n}\, dS = -\iiint_S \nabla\cdot\rho\mathbf{q}\, dv$$

$$= \frac{\partial}{\partial t}\iiint_S \rho\, dv = \iiint_S \frac{\partial \rho}{\partial t}\, dv.$$

because $m = \iiint_S \rho dv.$

As this relation holds whatever surface is taken we may equate the integrands to each other,

$$\frac{\partial \rho}{\partial t} + \nabla\cdot\rho\mathbf{q} = 0. \tag{269}$$

In Cartesian this is

$$\frac{\partial \rho}{\partial t} + \frac{\partial}{\partial x}\rho u + \frac{\partial}{\partial y}\rho v + \frac{\partial}{\partial z}\rho w = 0.$$

This equation is called the *equation of continuity.* It states that matter is neither created nor destroyed at any point in the fluid.

It is convenient here to employ a special notation to be used when we follow the fluid in its motion as distinct from considering the fluid as it passes by a fixed region in space. For example, the rate change of density of a *definite portion* of the fluid as it is followed in its motion, symbolized by the special notation $\frac{D\rho}{Dt}$, is equal to the rate of change of its density $\frac{\partial\rho}{\partial t}$ observed as it goes by a fixed point in space plus the rate of change $\mathbf{q}\cdot\nabla\rho$, due to its velocity $\mathbf{q}$, so that

$$\frac{D\rho}{Dt}=\frac{\partial\rho}{\partial t}+\mathbf{q}\cdot\nabla\rho. \qquad (270)$$

In fact whatever point-function is placed in the parenthesis

$$\frac{D}{Dt}(\ )=\frac{\partial}{\partial t}(\ )+\mathbf{q}\cdot\nabla(\ )$$

This corresponds in the Cartesian notation to

$$\frac{D\rho}{Dt}=\frac{\partial\rho}{\partial t}+\frac{\partial\rho}{\partial x}u+\frac{\partial\rho}{\partial y}v+\frac{\partial\rho}{\partial z}w$$

$$=\frac{\partial\rho}{\partial t}+\frac{\partial\rho}{\partial x}\frac{dx}{dt}+\frac{\partial\rho}{\partial y}\frac{dy}{dt}+\frac{\partial\rho}{\partial z}\frac{dz}{dt}.$$

We may write the equation of continuity in a slightly different form by (128) and using (270),

$$\frac{\partial\rho}{\partial t}+\nabla\cdot(\rho\mathbf{q})=\frac{\partial\rho}{\partial t}+\rho\nabla\cdot\mathbf{q}+\mathbf{q}\cdot\nabla\rho$$

so that

$$\frac{D\rho}{Dt}+\rho\nabla\cdot\mathbf{q}=0. \qquad (271)$$

In either form, if the fluid is *incompressible*, ρ does not vary either with time or with position, and hence

$$\nabla\cdot\mathbf{q}=0=\text{div}\,\mathbf{q}, \qquad (272)$$

which shows that $\mathbf{q}$ is then a solenoidal vector and its stream lines form closed curves or end at infinity, just as a solenoidal distribution of electrical force acts. In fact the two theories of Hydrodynamics of incompressible fluids and of Electricity are identical.

Consider now a small surface always containing the *same* fluid of volume v. This surface may be distorted as it moves with the liquid, but it is supposed to be always made up of the same small portions of the fluid with which it started. For such a surface, evidently the mass is constant, or

$$\frac{Dm}{Dt} = 0 = \frac{D\rho v}{Dt} = \rho\frac{Dv}{Dt} + v\frac{D\rho}{Dt},$$

so that

$$\frac{\frac{Dv}{Dt}}{v} = -\frac{\frac{D\rho}{Dt}}{\rho} = \nabla\cdot\mathbf{q} \text{ by (271).} \tag{273}$$

We may then interpret $\nabla\cdot\mathbf{q}$ as the fractional decrease of density per unit of time, or as the rate of increase of volume per unit volume, or as the time rate of dilatation, a divergence. Equation (272) follows also from (273) if ρ is constant.

Euler's Equations of Motion of a Fluid. Consider the forces acting upon a definite mass of the fluid enclosed in the surface S.

Let F per unit mass or ρ F per unit volume be the external force function, and let p be the pressure function acting normally over the enclosing surface and along the inwardly drawn normal. By Newton's law the rate of increase of momentum $(\Sigma\rho\,\mathbf{q}\,dv)$ of the fluid is equal to the applied forces F acting directly on the mass of the fluid and to the forces $(\Sigma p\,dS)$ resulting from the pressures acting on the surrounding surface, or

$$\frac{D}{Dt}\iiint_{\text{vol}}\rho\,\mathbf{q}\,dv = \iiint_{\text{vol}}\rho\,\mathbf{F}\,dv + \iint_S p\,\mathbf{n}\,dS,$$

or

$$\iiint\frac{D}{Dt}(\rho\,\mathbf{q}\,dv) = \iiint(\rho\,\mathbf{F} - \nabla p)\,dv.^*$$

Now

$$\frac{D}{Dt}(\rho\,\mathbf{q}\,dv) = \frac{D\mathbf{q}}{Dt}\rho\,dv + \mathbf{q}\frac{D(\rho\,dv)}{Dt},$$

and the last term vanishes as the **mass** remains constant throughout the motion, so that the integral becomes

$$\iiint\left(\frac{D\mathbf{q}}{Dt}\rho\right)dv = \iiint(\rho\,\mathbf{F} - \nabla p)\,dv. \tag{274}$$

* See note to § 52, p. 252, for transformation of last term by divergence theorem.

As this is true for any volume whatever, the integrands are equal and we have, using (270),

$$\rho \frac{D\mathbf{q}}{Dt} = \rho \frac{\partial \mathbf{q}}{\partial t} + \rho\, \mathbf{q}\cdot\nabla\mathbf{q} = \rho\mathbf{F} - \nabla p. \qquad (275)$$

This is Euler's equation of motion which, in connection with (268), (269) and (270), forms the basis of theoretical hydrodynamics.

99. Transformation of the Equation of Motion. If we divide (275) by ρ and employ the identity (129)

$$\mathbf{q}\cdot\nabla_1\mathbf{q}_1 = \nabla_1(\mathbf{q}_1\cdot\mathbf{q}) - \mathbf{q}\times(\nabla_1\times\mathbf{q}_1),$$

the subscripts indicating precisely on what the ∇ acts, or

$$\mathbf{q}\cdot\nabla\mathbf{q} = \tfrac{1}{2}\,\nabla\mathbf{q}^2 - \mathbf{q}\times(\nabla\times\mathbf{q}),$$

we may transform this equation into

$$\frac{\partial \mathbf{q}}{\partial t} - \mathbf{q}\times(\nabla\times\mathbf{q}) = \mathbf{F} - \frac{\nabla p}{\rho} - \frac{1}{2}\nabla\mathbf{q}^2. \qquad (276)$$

If the externally applied forces have a potential, V, for instance forces due to gravity or any other conservative system of forces, then

$$\mathbf{F} = -\nabla V.$$

If the pressure p at any point depends only upon the density ρ, we may define a quantity P such that

$$\frac{\nabla p}{\rho} = \nabla P, \text{ or } P = \int \frac{dp}{\rho},$$

so that our equation becomes

$$\frac{\partial \mathbf{q}}{\partial t} - \mathbf{q}\times \operatorname{curl} \mathbf{q} = -\nabla\left(V + P + \frac{1}{2}\,\mathbf{q}^2\right) \qquad (277)$$

$$= \nabla U,$$

where

$$U = -\,(V + P + \tfrac{1}{2}\,\mathbf{q}^2).$$

Referring back to equation (126) where it was shown that

$$\operatorname{curl} \mathbf{q} = 2\,\boldsymbol{\omega},$$

where $\boldsymbol{\omega}$ is the vorticity or angular velocity of rotation of the fluid at the point considered, (277) becomes

$$\frac{\partial \mathbf{q}}{\partial t} - 2\,\mathbf{q}\times\boldsymbol{\omega} = -\,\nabla\left(V + P + \frac{1}{2}\mathbf{q}^2\right).$$

100. Steady Motion. Definition. Steady motion is one in which $\mathbf{F}$, $\mathbf{q}$, p, and ρ are independent of the *time.* If such is the case and $\boldsymbol{\omega} = 0$, that is, if the motion is non-vortical,

$$-\nabla U = \nabla\left(V + P + \tfrac{1}{2}\,\mathbf{q}^2\right) = 0,$$

or integrating $\qquad V + P + \frac{1}{2}\,\mathbf{q}^2 = \text{const.}$

If ρ is constant,

$$P = \frac{p}{\rho},$$

and if there are no applied forces,

$$\mathbf{F} = 0$$

and hence $\qquad V = \text{const.},$

so that

$$\frac{p}{\rho} + \frac{\mathbf{q}^2}{2} = \text{const.} \tag{278}$$

In other words, where the pressure is great the velocity must be small, and where the velocity is great the pressure is small. For example, in a constricted pipe the pressure is least at the constriction where the velocity of the incompressible fluid necessarily is the greatest. Air pumps and water meters are constructed on this principle.

101. Vortex Motion. Theorem of Helmholtz. Take the curl, or apply $\nabla\times$ to

$$\frac{\partial \mathbf{q}}{\partial t} + 2\,\boldsymbol{\omega}\times\mathbf{q} = -\nabla\left(V + P + \frac{\mathbf{q}^2}{2}\right),$$

giving

$$\text{curl}\,\frac{\partial \mathbf{q}}{\partial t} + 2\,\nabla\times(\boldsymbol{\omega}\times\mathbf{q}) = 0, \qquad \text{as curl grad} \equiv 0$$

or $\qquad \dfrac{\partial}{\partial t}\,(\text{curl}\,\mathbf{q}) + 2\,(\boldsymbol{\omega}\nabla\cdot\mathbf{q} + \mathbf{q}\cdot\nabla\boldsymbol{\omega} - \mathbf{q}\nabla\cdot\boldsymbol{\omega} - \boldsymbol{\omega}\cdot\nabla\mathbf{q}) = 0.$

Remembering that

$$\text{curl}\,\mathbf{q} = 2\,\boldsymbol{\omega}; \quad \text{that} \quad \nabla\cdot\boldsymbol{\omega} = \tfrac{1}{2}\,\nabla\cdot\nabla\times(\mathbf{q}) = 0,$$

as $\boldsymbol{\omega}$ is a solenoidal vector, *i.e.*, $\boldsymbol{\omega} = \frac{1}{2}$ curl $\mathbf{q}$, and that (270)

$$\frac{\partial \boldsymbol{\omega}}{\partial t} + \mathbf{q}\cdot\nabla\boldsymbol{\omega} = \frac{D\boldsymbol{\omega}}{Dt},$$

we have
$$\frac{D\boldsymbol{\omega}}{Dt} + \boldsymbol{\omega}\,\nabla\cdot\mathbf{q} - \boldsymbol{\omega}\cdot\nabla\mathbf{q} = 0.$$

This transforms into

$$\frac{D}{Dt}\left(\frac{\boldsymbol{\omega}}{\rho}\right) = \frac{\boldsymbol{\omega}}{\rho}\cdot\nabla\mathbf{q}, \qquad (279)$$

because identically, using (273)

$$\nabla\cdot\mathbf{q} = -\frac{1}{\rho}\frac{D\rho}{Dt},$$

$$\frac{D\boldsymbol{\omega}}{Dt} + \boldsymbol{\omega}\,\nabla\cdot\mathbf{q} = \frac{D\boldsymbol{\omega}}{Dt} - \frac{\boldsymbol{\omega}}{\rho}\frac{D\rho}{Dt} \equiv \rho\,\frac{D}{Dt}\left(\frac{\boldsymbol{\omega}}{\rho}\right).$$

Hence, also, differentiating (279) again

$$\frac{D^2}{Dt^2}\left(\frac{\boldsymbol{\omega}}{\rho}\right) = \left(\frac{D}{Dt}\frac{\boldsymbol{\omega}}{\rho}\right)\cdot\nabla\mathbf{q} + \frac{\boldsymbol{\omega}}{\rho}\cdot\frac{D}{Dt}\nabla\mathbf{q}, \qquad (280)$$

so that if $\boldsymbol{\omega}$ ever vanishes (279) and (280) likewise vanish, and similarly all the successive derivatives may be shown to vanish. Hence if $\boldsymbol{\omega}$ is ever zero it will always remain zero by Taylor's theorem, because all of its derivatives vanish at a certain instant. This theorem due to Helmholtz says that if no vorticity exists in any incompressible, frictionless fluid at any time it is impossible to produce any by means of a conservative system of forces, and the motion will remain forever non-vortical.

* If the ether be considered to be a frictionless medium, then a vortex once set up in it would be indestructible; and conversely, if no vortices existed, it would be impossible to create any. It is conceivable, however, that some "Cataclysm" might have rendered the ether temporarily viscous to some extent. By this we mean that it is conceivable, for example, that under extraordinary conditions say of temperature the ether might acquire unusual properties, in which case, if it became frictionless again after vortical motion had been produced in it while in this state, such vortical motion would persist forever. This speculation is of interest in connection with the vortex-atom theory of matter.

102. Circulation. The circulation along any path in a fluid is defined as the line-integral of the velocity along that path. If ϕ_{AB} denote the circulation along the path AB, by definition

$$\phi_{AB} = \int_A^B \mathbf{q} \cdot d\mathbf{r}. \tag{281}$$

If the path is a closed one, we may express the circulation around it as a surface integral over any cap bounded by it, by means of Stokes' Theorem, for

$$\int_{\text{C}} \mathbf{q} \cdot d\mathbf{r} = \iint_{\text{cap}} \mathbf{n} \cdot \nabla \times \mathbf{q} \, dS = 2 \iint_{\text{cap}} \mathbf{n} \cdot \boldsymbol{\omega} \, dS,$$

where

$$2\,\boldsymbol{\omega} = \nabla \times \mathbf{q} \quad \text{by (126).}$$

Roughly speaking this equation says, if there is a preponderance of motion of a liquid in one direction or the other

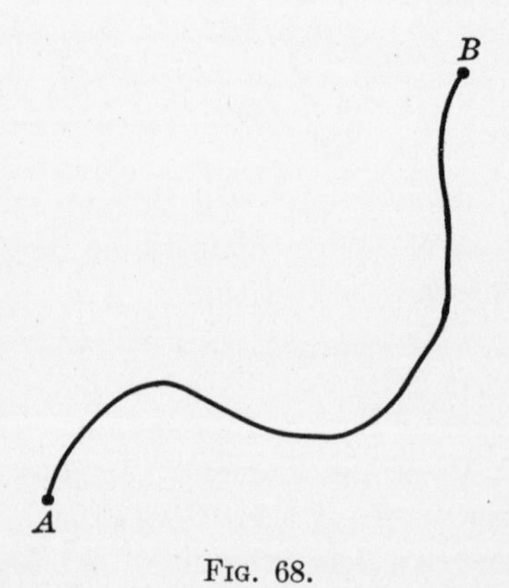

Fig. 68.

around any closed path, that the liquid inside of the closed path must be rotating.

Consider a tube made up of the lines of the vector $\boldsymbol{\omega}$, and consider a portion of it bounded by two caps S_1 and S_2.

Apply the divergence theorem to this closed surface S_1, S_2 and sides. Remembering that $\boldsymbol{\omega}$ is solenoidal, we have

$$\iint_S \mathbf{n} \cdot \boldsymbol{\omega} \, dS = \iiint_{\text{vol } S} \nabla \cdot \boldsymbol{\omega} \, dv = 0.$$

As the sides contribute nothing to the surface integral there must be as much flux of $\boldsymbol{\omega}$ inward at S_1 as there is outward at S_2, or the flux is constant throughout the tube. If

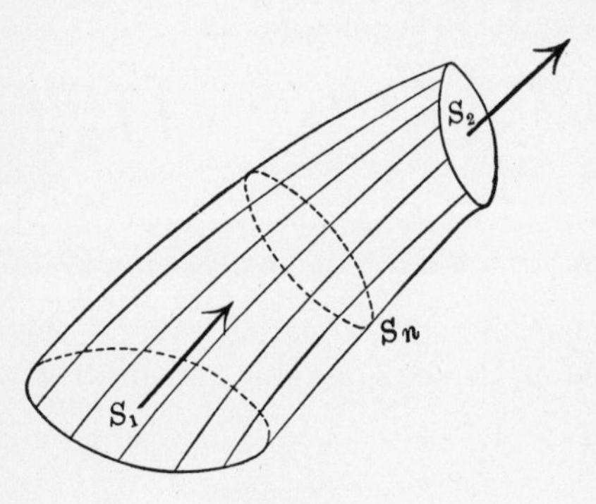

FIG. 69.

this tube be chosen very small it is called a vortex filament, and if the section of such a filament be denoted by s the above result expresses the fact that

$$s \, \mathbf{n} \cdot \boldsymbol{\omega} = \text{const.}, \tag{282}$$

where $\mathbf{n}$ is the normal to the cross-section. This product is called the strength of the filament. It shows that if $\boldsymbol{\omega}$ is finite, or, in other words, if there does exist any vorticity, s cannot vanish, hence a filament cannot end anywhere *in* the fluid. Such filaments must then either form closed curves or end in the surface of the liquid or at infinity. All vortices, then, form closed curves in the fluid or else end in the surface. This also follows from the fact that $\nabla \cdot \boldsymbol{\omega} = 0$, that is, $\boldsymbol{\omega}$ is a solenoidal vector.

103. Velocity-Potential. If $\boldsymbol{\omega}$ is zero everywhere, the circulation around any closed curve is zero, hence the circulation from any point A to any other point B is independent of the path. In this case $\mathbf{q}\cdot d\mathbf{r}$ is a perfect differential; that is, it is of the form

$$d\mathbf{r}\cdot\mathbf{q} = d\phi = d\mathbf{r}\cdot\nabla\phi,$$

so that
$$\mathbf{q} = \nabla\phi. \tag{283}$$

The velocity $\mathbf{q}$ is thus derivable from the function ϕ in the same way (except for sign), that the force is derivable from the ordinary potential. Accordingly ϕ is called the Velocity-Potential, and is a scalar point-function of the space occupied by the fluid. All the results of the theory of potential are therefore directly applicable to the function ϕ.

Production of a Vortex Impossible in a Frictionless Fluid. Let us find the time-rate of variation of the circulation along any path, assuming a velocity-potential to exist. This path is made up of certain elements of the fluid which are to be followed in their motion, however distorted the path may become. Differentiating (281),

$$\frac{D\phi}{Dt} = \frac{D}{Dt}\int_A^B \mathbf{q}\cdot d\mathbf{r} = \int_A^B \frac{D\mathbf{q}}{Dt}\cdot d\mathbf{r} + \int_A^B \mathbf{q}\cdot\frac{D}{Dt}\, d\mathbf{r}.$$

Since the velocity-potential exists,

$$\text{curl}\,\mathbf{q} = 2\,\boldsymbol{\omega} = 0,$$

and the equation of motion (275) becomes, if $W \equiv (-V + P)$

$$\frac{D\mathbf{q}}{Dt} = \nabla W, \tag{284}$$

so that
$$\frac{D\mathbf{q}}{Dt}\cdot d\mathbf{r} = d\mathbf{r}\cdot\nabla W = dW,$$

also
$$\mathbf{q}\cdot\frac{D}{Dt}\, d\mathbf{r} = \mathbf{q}\cdot d\,\frac{D\mathbf{r}}{Dt} = \mathbf{q}\cdot d\mathbf{q} = d\left(\frac{\mathbf{q}^2}{2}\right)$$

and hence
$$\frac{D\phi}{Dt} = \int_A^B d\left(W + \frac{\mathbf{q}^2}{2}\right) = \left[W + \frac{\mathbf{q}^2}{2}\right]_A^B. \tag{285}$$

If the path is a closed one,

$$\left[W + \frac{\mathbf{q}^2}{2}\right]_A^B = 0,$$

as W and $\frac{q^2}{2}$ are scalar point-functions of position and have identical values at the limits, so that finally

$$\frac{D\phi}{Dt} = 0. \tag{286}$$

Equation (286) then states that the circulation around any closed curve, formed of a chain of particles of the fluid, cannot change as these particles are carried about by the liquid. As we have assumed the circulation to be zero at the beginning it remains so forever, or, in other words, it is impossible to create vorticity in frictionless fluid by means of a *conservative* system of forces. Also, as it is impossible to conceive how any system of forces could act on a *frictionless* fluid in a *non*-conservative manner, it follows that it is impossible to create vorticity in any manner in a frictionless medium.

It was from these peculiar properties of vortices in a frictionless fluid, discovered by Helmholtz, that **Lord Kelvin** was led to his **Vortex Atom Theory of Matter.**

PROBLEMS AND EXERCISES

1. Show that the center of gravity of a system of particles, and hence of any body, continues to move uniformly in a straight line when no impressed forces act upon the system.

Find the equation of the path.

2. Show that the total momentum of a system of particles, and hence of any body, remains constant as long as there are no applied forces.

3. Prove that a system of forces acting along and represented by the sides of a plane polygon taken in order is equivalent to a couple whose moment is represented by twice the area of the polygon. Extend this to forces acting along a closed plane curve.

4. By means of the theorem (202)

$$\int_{\mathfrak{C}} \mathbf{q} \times d\mathbf{r} = \iint_{\text{cap}} [\mathbf{n}\nabla \cdot \mathbf{q} - \nabla(\mathbf{q} \cdot \mathbf{n})]\, dS$$

show that if forces equal in magnitude act everywhere along the tangents to a plane contour, that the moment of these forces about any point is measured by twice the area of the contour.

5. If a rigid body has a velocity of translation $\mathbf{q}_t$ and an angular velocity of rotation $\boldsymbol{\omega}$, the velocity $\mathbf{q}$ at any instant of a point $\mathbf{r}$ in the body may be represented by

$$\mathbf{q} = \mathbf{q}_t + \boldsymbol{\omega} \times \mathbf{r}. \qquad \text{see § 22.}$$

Show that if $\mathbf{q}_t$ and $\boldsymbol{\omega}$ are constants the path of any point in the body is a circular helix described with uniform velocity, and find its equation.

6. Show that two equal rotations in opposite directions about two parallel axes produce a motion perpendicular to the plane of the two axes.

7. The motion of a point in a plane being given, refer it to
(a) fixed rectangular vectors in the plane;
(b) rectangular vectors in the plane, revolving uniformly about a fixed point.

Translate into Cartesian in both cases.

8. Prove that the central axis of two forces $\mathbf{F}_1$ and $\mathbf{F}_2$ intersects the shortest distance between their lines of action and divides it in the ratio

$$F_2 (F_2 + F_1 \cos \theta) : F_1 (F_1 + F_2 \cos \theta),$$

θ being the angle between their directions. Also prove that the moment of the principal couple is

$$\frac{cF_1F_2 \sin \theta}{\sqrt{F_1^2 + F_2^2 + 2\, F_1F_2 \cos \theta}}.$$

9. Show that

$$\iint_S \mathbf{n} \times \mathbf{F}\, dS = \iiint_{\text{vol}} \nabla \times \mathbf{F}\, dv.$$

What conclusion in Hydrodynamics does this theorem lead to?

10. Express the following equation in vector notation:

$$\iint\left[\left\{\frac{\partial w}{\partial y}-\frac{\partial v}{\partial z}\right\}\cos(nx)+\left\{\frac{\partial u}{\partial z}-\frac{\partial w}{\partial x}\right\}\cos(ny)+\left\{\frac{\partial v}{\partial x}-\frac{\partial u}{\partial y}\right\}\cos(nz)\right]dS.$$

Ans. $\iint (\nabla\times\mathbf{q})\cdot\mathbf{n}\, dS.$

11. Express the following equations in vector notation:

$$\frac{\partial u}{\partial t}+2(w\eta-v\zeta)=\frac{\partial U}{\partial x}-\frac{1}{\rho}\frac{\partial p}{\partial x},$$

$$\frac{\partial v}{\partial t}+2(u\zeta-w\xi)=\frac{\partial U}{\partial y}-\frac{1}{\rho}\frac{\partial p}{\partial y},$$

$$\frac{\partial w}{\partial t}+2(v\xi-u\eta)=\frac{\partial U}{\partial z}-\frac{1}{\rho}\frac{\partial p}{\partial z}.$$

Let $\mathbf{q}$ have components u, v and w, and $\boldsymbol{\omega}$ have components η, ζ and ξ.

Ans. $\frac{\partial\mathbf{q}}{\partial t}+2\boldsymbol{\omega}\times\mathbf{q}=\nabla U-\frac{1}{\rho}\nabla p.$

12. Express in vector notation the following equations which occur in the theory of Elasticity:

$$\rho\frac{\partial^2 u}{\partial t^2}=(\lambda+\mu)\frac{\partial\sigma}{\partial x}+\mu\left(\frac{\partial^2 u}{\partial x^2}+\frac{\partial^2 u}{\partial y^2}+\frac{\partial^2 u}{\partial z^2}\right),$$

$$\rho\frac{\partial^2 v}{\partial t^2}=(\lambda+\mu)\frac{\partial\sigma}{\partial y}+\mu\left(\frac{\partial^2 v}{\partial x^2}+\frac{\partial^2 v}{\partial y^2}+\frac{\partial^2 v}{\partial z^2}\right),$$

$$\rho\frac{\partial^2 w}{\partial t^2}=(\lambda+\mu)\frac{\partial\sigma}{\partial z}+\mu\left(\frac{\partial^2 w}{\partial x^2}+\frac{\partial^2 w}{\partial y^2}+\frac{\partial^2 w}{\partial z^2}\right),$$

where u, v, w are the components of a vector **q**, where σ is a scalar variable, and where ρ, λ, and μ are constants.

Ans. $\rho\frac{\partial^2\mathbf{q}}{\partial t^2}=(\lambda+\mu)\nabla\sigma+\mu\nabla^2\mathbf{q}.$

13. Express the following equations in vector notation:

$$\frac{\partial u}{\partial t} + u\frac{\partial u}{\partial x} + v\frac{\partial u}{\partial y} + w\frac{\partial u}{\partial z} - \frac{\nu}{3}\frac{\partial \sigma}{\partial x} - \nu\left(\frac{\partial^2 u}{\partial x^2} + \frac{\partial^2 u}{\partial y^2} + \frac{\partial^2 u}{\partial z^2}\right) = X - \frac{1}{\rho}\frac{\partial p}{\partial x},$$

$$\frac{\partial v}{\partial t} + u\frac{\partial v}{\partial x} + v\frac{\partial v}{\partial y} + w\frac{\partial v}{\partial z} - \frac{\nu}{3}\frac{\partial \sigma}{\partial y} - \nu\left(\frac{\partial^2 v}{\partial x^2} + \frac{\partial^2 v}{\partial y^2} + \frac{\partial^2 v}{\partial z^2}\right) = Y - \frac{1}{\rho}\frac{\partial p}{\partial y},$$

$$\frac{\partial w}{\partial t} + u\frac{\partial w}{\partial x} + v\frac{\partial w}{\partial y} + w\frac{\partial w}{\partial z} - \frac{\nu}{3}\frac{\partial \sigma}{\partial z} - \nu\left(\frac{\partial^2 w}{\partial x^2} + \frac{\partial^2 w}{\partial y^2} + \frac{\partial^2 w}{\partial z^2}\right) = Z - \frac{1}{\rho}\frac{\partial p}{\partial z},$$

where u, v, and w are components of the vector $\mathbf{q}$, where X, Y, and Z are components of the vector $\mathbf{F}$, where p and σ are scalar variables, and where ν is a scalar constant.

Ans. $$\frac{\partial \mathbf{q}}{\partial t} + \mathbf{q}\cdot\nabla\mathbf{q} - \frac{\nu}{3}\nabla\sigma - \nu\nabla^2\mathbf{q} = \mathbf{F} - \frac{1}{\rho}\nabla p.$$

APPENDIX.

NOTATION AND FORMULÆ.

The Various Notations.

Whenever new quantities are introduced it is well to have as simple and as convenient a notation as possible. The notation devised by the late **Professor Willard Gibbs** seemed to us, after much thought on the matter, to be the simplest and most symmetrical of any of the *existing* kinds.

Hamilton, the inventor of quaternions, used the letters S and V for the scalar and vector products respectively of the vectors that followed them; thus

$$S\mathbf{ab} \text{ and } V\mathbf{ab}$$

represented respectively the scalar and vector products of the vectors **a** and **b**.

The letter T, standing for tensor, represented the magnitude of the vector following it. This notation has many advantages, but after deliberation it was discarded.

Oliver Heaviside, the English electrician, used a notation similar to Hamilton's, but rendered it unsymmetrical by discarding the S for the scalar product while retaining the V for a vector product. This seemed to us to be a step backward, although he was followed in its use by Föppl and Bucherer in Germany and by others.

The disciples of **Grassmann,** who had devised a notation of his own, adapted it to the analysis of vectors, and at the present time the resulting notation has a number of adherents in Germany and elsewhere. Our main objection to it is that it uses different kinds of parentheses to distinguish the two products, thus preventing the use of these parentheses for other purposes. It is also quite cumbersome and takes much longer to write than any of the other systems, besides there being a liability of error due to the fact that all parentheses necessarily look somewhat alike.

Gibbs, on the other hand, puts the distinguishing product mark *between* the two vectors instead of in front or around them.

This is essentially a symmetrical notation, and to our mind and to many others the best. The two symbols used to indicate the " variety " of product are the dot (·) and the cross (×). In order to avoid any confusion with the ordinary dot and cross used for ordinary products, and a necessity in any analysis, we have ventured to use a special dot and a special cross. That is, the dot is above the writing line and the cross is a small one and when used is placed in the same position as the dot. Thus

$$\mathbf{a} \cdot \mathbf{b} \text{ and } \mathbf{a} \times \mathbf{b}$$

are the scalar and vector products of the vectors **a** and **b** respectively.

They are easy to write, easily distinguished and connected with the *idea* of a product. They do not interfere with parentheses, neither do they render the use of an ordinary dot (.) or a cross (×) undesirable nor ambiguous in other parts of the work. They are symmetrically placed.

Comparison of Notations.

A Few Examples of Formulæ in the four systems of notation will render the foregoing clear to the student. We shall give Hamilton's notation the benefit of our bold-faced type and avoid the wholesale use of Greek letters which were employed by him to represent vectors.

The formulæ are in the order:

1. Gibbs' Notation.
2. Hamilton's Notation.
3. Heaviside's Notation.
4. Gans' Notation. (Grassmannian.)

1. a or $\mathbf{a}_0$.
2. $T\mathbf{a}$.
3. a.
4. $|a|$.

1. $\mathbf{a}\cdot\mathbf{b} = \mathbf{b}\cdot\mathbf{a} = ab \cos (\mathbf{ab})$.
2. $\mathbf{Sab} = \mathbf{Sba} = - T\mathbf{a}\, T\mathbf{b} \cos (\mathbf{ab})$.
3. $\mathbf{ab} = \mathbf{ba} = ab \cos (\mathbf{ab})$.
4. $(ab) = (ba) = |a||b| \cos (ab)$.

1. $\mathbf{a}\times\mathbf{b} = - \mathbf{b}\times\mathbf{a} = \epsilon\, ab \sin (\mathbf{ab})$.
2. $\mathbf{Vab} = -\mathbf{Vba} = \epsilon\, T\mathbf{a}\, T\mathbf{b} \sin (\mathbf{ab})$.
3. $\mathbf{Vab} = -\mathbf{Vba} = \epsilon\, ab \sin (\mathbf{ab})$.
4. $[ab] = - [ba] = \epsilon\, |a||b| \sin (ab)$.

1. $\mathbf{a}\cdot(\mathbf{b} + \mathbf{c}) = \mathbf{a}\cdot\mathbf{b} + \mathbf{a}\cdot\mathbf{c}$. $\quad \mathbf{a}\cdot\mathbf{b}\times\mathbf{c} = \mathbf{b}\cdot\mathbf{c}\times\mathbf{a}$.
2. $\mathbf{Sa}(\mathbf{b} + \mathbf{c}) = \mathbf{Sab} + \mathbf{Sac}$. $\quad \mathbf{SaVbc} = \mathbf{SbVca}$.
3. $\mathbf{a}(\mathbf{b} + \mathbf{c}) = \mathbf{ab} + \mathbf{ac}$. $\quad \mathbf{aVbc} = \mathbf{bVca}$.
4. $(a, b + c) = (ab) + (ac)$. $\quad (a[bc]) = (b[ca])$.

1. $\mathbf{a}\times(\mathbf{b}\times\mathbf{c}) = \mathbf{b}\,\mathbf{a}\cdot\mathbf{c} - \mathbf{c}\,\mathbf{a}\cdot\mathbf{b}$.
2. $\mathbf{VaVbc} = \mathbf{cSab} - \mathbf{bSac}$.
3. $\mathbf{VaVbc} = \mathbf{b\,ac} - \mathbf{c\,ab}$.
4. $[a[bc]] = b(ac) - c(ab)$.

1. $$\mathbf{r} = \frac{\mathbf{r}\cdot\mathbf{b}\times\mathbf{c}}{[\mathbf{abc}]}\mathbf{a} + \frac{\mathbf{r}\cdot\mathbf{c}\times\mathbf{a}}{[\mathbf{abc}]}\mathbf{b} + \frac{\mathbf{r}\cdot\mathbf{a}\times\mathbf{b}}{[\mathbf{abc}]}\mathbf{c}.$$
2. $$\mathbf{r} = \frac{\mathbf{SrVbc}}{\mathbf{SaVbc}}\mathbf{a} + \frac{\mathbf{SrVca}}{\mathbf{SaVbc}}\mathbf{b} + \frac{\mathbf{SrVab}}{\mathbf{SaVbc}}\mathbf{c}.$$
3. $$\mathbf{r} = \frac{\mathbf{rVbc}}{\mathbf{aVbc}}\mathbf{a} + \frac{\mathbf{rVca}}{\mathbf{aVbc}}\mathbf{b} + \frac{\mathbf{rVab}}{\mathbf{aVbc}}\mathbf{c}.$$
4. $$r = \frac{(r[bc])}{(a[bc])}a + \frac{(r[ca])}{(a[bc])}b + \frac{(r[ab])}{(a[bc])}c.$$

These examples are sufficient to show the *characteristics* of the various notations.

Notation of this Book.

It seems to be the consensus of opinion that vectors are best represented by single letters printed in some sort of **bold-faced** type. If this is not done, in order to distinguish a vector from a scalar we are obliged to employ Greek or other special alphabets and thus deprive ourselves of the convenience of using that alphabet if desired, and also we are prevented from using any other letter as a vector.

The Magnitude of a vector is represented by the same letter as the vector itself but in ordinary or *italic* type.

A Unit Vector parallel to any vector is represented by that vector with the subscript unity. Thus the vector

$$\mathbf{a}$$

has a magnitude a

and a direction $\mathbf{a}_1$,

so that we may write $\mathbf{a} = a\,\mathbf{a}_1$.

Sometimes the subscript zero to a vector, or particularly to a vector-expression, will mean that its magnitude alone is expressed. Thus

$$(\mathbf{a}\times\mathbf{b})_0$$

denotes the magnitude of $\mathbf{a}\times\mathbf{b}$.

In order to connect the *Analysis of Vectors* with *Cartesian Analysis* it is necessary to relate the vector **a** to its three components along the three Cartesian axes. We shall denote these components by adding the subscripts 1, 2, and 3 to the italic letter a.

As mutually perpendicular axes are by far the most important of any, three unit vectors

i, **j**, and **k**

have been *universally* adopted to represent their directions in space.

The vector **a** then is made up of

a vector along **i**, of length a_1,
a vector along **j**, of length a_2,
and a vector along **k**, of length a_3,

so that

$$\mathbf{a} = a\,\mathbf{a}_1 = a_1\mathbf{i} + a_2\mathbf{j} + a_3\mathbf{k}.$$

There can be no confusion between the letters $\mathbf{a}_1$ and a_1, for obvious reasons.

A Voluntary Exception to this convention is in the case of the radius vector **r** to any point from the origin. Its components will be denoted by x, y, and z instead of r_1, r_2, and r_3 in order to approach more closely to the usual Cartesian character of the work when translated into that notation. For this reason we write

$$\mathbf{r} = x\,\mathbf{i} + y\,\mathbf{j} + z\,\mathbf{k},$$

and for similar reasons when desirable,

$$\mathbf{F} = X\,\mathbf{i} + Y\,\mathbf{j} + Z\,\mathbf{k},$$
$$\mathbf{q} = u\,\mathbf{i} + v\,\mathbf{j} + w\,\mathbf{k},$$
$$\boldsymbol{\omega} = \xi\,\mathbf{i} + \eta\,\mathbf{j} + \zeta\,\mathbf{k}.$$

The Unit Tangent along and the **Unit Normal** to a curve or surface are denoted by the unit vectors **t** and **n**. Since the components of a unit vector are its direction cosines in

$$\mathbf{t} = t_1\mathbf{i} + t_2\mathbf{j} + t_3\mathbf{k}$$

and

$$\mathbf{n} = n_1\mathbf{i} + n_2\mathbf{j} + n_3\mathbf{k},$$

t_1, t_2, and t_3 and n_1, n_2, and n_3 are the direction cosines of the tangent and normal respectively.

The Scalar or Dot Product of two vectors is represented by placing a dot between them thus:

$$\mathbf{a}\cdot\mathbf{b} = \mathbf{b}\cdot\mathbf{a} = ab\cos(\mathbf{ab}),$$

where cos (**ab**) is the notation used for the cosine of the angle included by the positive directions of **a** and **b**.

The Vector or Cross Product of two vectors is represented by placing a special cross between them thus:

$$\mathbf{a}\times\mathbf{b} = -\,\mathbf{b}\times\mathbf{a} = \boldsymbol{\epsilon}\, ab \sin(\mathbf{ab}).$$

The unit vector $\boldsymbol{\epsilon}$ is a vector perpendicular to the two vectors in the product and is taken in such a " sense " that as you turn **a** into **b**, $\boldsymbol{\epsilon}$ points in the direction a cork-screw would advance if so rotated.

Then $$(\mathbf{a}\times\mathbf{b})_1 = \boldsymbol{\epsilon}$$

and $$(\mathbf{a}\times\mathbf{b})_0 = ab \sin(\mathbf{ab}).$$

A Scalar Point-Function is represented by writing the functional symbol in italic. Thus

$$V = f(\mathbf{r})$$

means that V is a scalar point-function of the radius vector **r**. That is, for every value of **r**, V has a determinate *magnitude.* This is equivalent to *one* Cartesian equation.

A Vector Point-Function is represented by writing the functional symbol in bold-faced type. Thus

$$\mathbf{F} = \mathbf{f}(\mathbf{r})$$

means that **F** is a vector point-function of the radius vector **r**. That is, for every value of **r**, **F** has a determinate *magnitude* and *direction.* This is equivalent to *three* Cartesian equations.

A Linear Vector Function, in particular, is represented by the special symbols

$$\boldsymbol{\omega},\ \boldsymbol{\psi} \text{ or } \boldsymbol{\chi}.$$

Velocity and Angular Velocity are represented by the symbols respectively

$$\mathbf{q} \text{ and } \boldsymbol{\omega}.$$

The Scalar Potential Function is represented by the italics

$$V \text{ or } \Omega.$$

The Vector Potential is represented by the bold-faced

$$\mathbf{V}.$$

Electric or Magnetic Intensities or Forces in General are represented by

$$\mathbf{F} \text{ or } \mathbf{H}.$$

The Differential Vector Operator ∇ (read *del*) is equivalent to

$$\nabla(\;) \equiv \left(\mathbf{i}\frac{\partial}{\partial x} + \mathbf{j}\frac{\partial}{dy} + \mathbf{k}\frac{\partial}{\partial z}\right)(\;).$$

Besides obeying the laws obeyed by ordinary vectors, it is a differentiating operator, and the same care should be taken in its use and interpretation as should be taken with other differentiating operators.

As Professor Joly* puts it:

" Of course some little care is necessary when ∇ is expressed in the general form, but it is precisely of the same kind as the care required to distinguish between

$$\left(x^2\frac{\partial}{\partial x}\right)^2 \equiv \left(x^2\frac{\partial}{\partial x}\right)\left(x^2\frac{\partial}{\partial x}\right) = x^4\frac{\partial}{\partial x}\left(\frac{\partial}{\partial x}\right) + x^2\frac{\partial}{\partial x}\left(\frac{\partial}{\partial x}x^2\right)$$

and

$$x^4\left(\frac{\partial}{\partial x}\right)^2 = x^4\frac{\partial}{\partial x}\left(\frac{\partial}{\partial x}\right)."$$

Del sometimes differentiates partially. Generally a subscript attached to it, indicates the variable which it differentiates. Thus

$$\nabla_a \mathbf{a}\cdot\mathbf{b}$$

means that in the scalar product the vector **a** above is to be considered *variable.*

Sometimes the same process may be indicated by writing as a subscript to *the expression* the quantity which is to remain *constant* during the differentiation. Thus

$$\nabla(\mathbf{a}\cdot\mathbf{b})_b \text{ is the same as } \nabla_a(\mathbf{a}\cdot\mathbf{b}).$$

* Charles Jasper Joly, Manual of Quaternions, Macmillan & Co., 1905, Art. 57, p. 75.

Which notation is preferable will depend on whether all but one of the vectors are to be considered variable or whether all but one are constant, respectively.

Sometimes also the notation

$$\nabla'(\mathbf{a}'\cdot\mathbf{a} + \mathbf{a}'\cdot\mathbf{b})$$

is useful, the variables having the same accent as the ∇, being alone considered variable during the differentiation.

Gradient (grad), or vector of greatest slope of a scalar function V, is a vector normal to the level surfaces of the function V and is given by the operation of ∇ upon the function thus:

$$\nabla V = \text{gradient or slope of } V = \text{grad } V.$$

The Divergence of a vector function $\mathbf{F}$ is given by forming the scalar product of ∇ with the function thus

$$\nabla\cdot\mathbf{F} = \text{divergence of } \mathbf{F} = \text{div } \mathbf{F}.$$

The Curl of a vector function $\mathbf{F}$ is obtained by forming the vector product of ∇ with the function:

$$\nabla\times\mathbf{F} = \text{curl } \mathbf{F} = \text{rotation of } \mathbf{F} \text{ (rot } \mathbf{F}).$$

The more important formulæ of vector analysis are collected below for reference.

FORMULÆ.*

Vectors.

$$\begin{aligned} \mathbf{a} &= a\,\mathbf{a}_1 = \mathbf{a}_0\mathbf{a}_1 \\ &= a_1\mathbf{i} + a_2\mathbf{j} + a_3\mathbf{k}. \end{aligned} \qquad (1)$$

$$\mathbf{r} = x\,\mathbf{i} + y\,\mathbf{j} + z\,\mathbf{k}. \qquad (4)$$

$$\mathbf{F} = X\,\mathbf{i} + Y\,\mathbf{j} + Z\,\mathbf{k}.$$

$$\begin{aligned} \mathbf{q} &= q_1\mathbf{i} + q_2\mathbf{j} + q_3\,\mathbf{k} \\ &= u\,\mathbf{i} + v\,\mathbf{j} + w\,\mathbf{k}. \end{aligned}$$

$$\begin{aligned} \boldsymbol{\omega} &= \omega_1\mathbf{i} + \omega_2\mathbf{j} + \omega_3\,\mathbf{k} \\ &= \xi\,\mathbf{i} + \eta\;\mathbf{j} + \zeta\,\mathbf{k}. \end{aligned}$$

$$\Sigma\,\mathbf{a} = \mathbf{i}\,\Sigma a_1 + \mathbf{j}\,\Sigma a_2 + \mathbf{k}\,\Sigma a_3. \qquad (7)$$

$$\frac{1}{\mathbf{a}} = \mathbf{a}^{-1} = \frac{\mathbf{a}_1}{a} \qquad (2)$$

$$\frac{\mathbf{r}}{r} = \mathbf{r}_1 = \mathbf{i}\cos\alpha + \mathbf{j}\cos\beta + \mathbf{k}\cos\gamma, \qquad (9)$$

where α, β, γ are the direction angles of the radius vector **r**.

The equation of a straight line through the terminus of **b** and parallel to **a** is, s being a scalar variable,

$$\mathbf{r} = \mathbf{b} + s\,\mathbf{a}. \qquad (11)$$

If it passes through the origin,

$$\mathbf{r} = s\,\mathbf{a}. \qquad (10)$$

A line through the ends of **a** and **b**

$$\mathbf{r} = s\,\mathbf{a} + (1 - s)\,\mathbf{b}$$

or

$$\mathbf{r} = s\,\mathbf{b} + (1 - s)\,\mathbf{a}. \qquad (12)$$

The condition that three vectors, **a**, **b**, and **c**, should end in the same straight line is

$$\begin{aligned} x\,\mathbf{a} + y\,\mathbf{b} + z\,\mathbf{c} &= 0. \\ x + y + z &= 0. \end{aligned} \qquad (13)$$

* The numbering corresponds with that of the text.

The equation of a plane determined by the vectors **a** and **b**, and passing through the terminus of **c**, is

$$\mathbf{r} = \mathbf{c} + s\,\mathbf{a} + t\,\mathbf{b}. \tag{15}$$

The plane through the origin and parallel to **a** and **b** is

$$\mathbf{r} = s\,\mathbf{a} + t\,\mathbf{b}. \tag{14}$$

The equation of a plane passing through the three points **a**, **b**, and **c** is

$$\mathbf{r} = s\,\mathbf{a} + t\,\mathbf{b} + (1 - s - t)\,\mathbf{c}. \tag{16}$$

The condition that four vectors, **a**, **b**, **c**, and **d**, should end in the same plane is

$$\left.\begin{aligned} x\,\mathbf{a} + y\,\mathbf{b} + z\,\mathbf{c} + w\,\mathbf{d} &= 0. \\ x + y + z + w &= 0. \end{aligned}\right\} \tag{17}$$

The vector

$$\mathbf{r} = \frac{m\,\mathbf{a} + n\,\mathbf{b}}{m + n} \tag{18}$$

divides the line joining the points **a** and **b** in the ratio of m to n.

The vector to the center of gravity of the masses m, at points **a**, is

$$\bar{\mathbf{r}} = \frac{\Sigma m\,\mathbf{a}}{\Sigma m}. \tag{20}$$

If the relation

$$m_1\mathbf{a}_1 + m_2\mathbf{a}_2 + \ldots = 0 \tag{25}$$

is to be independent of the origin chosen, then

$$m_1 + m_2 + \ \ldots = 0.$$

Vector and Scalar Products.

Products of Two Vectors.

$$\mathbf{a}\cdot\mathbf{b} = ab\cos(\mathbf{ab}) = \mathbf{b}\cdot\mathbf{a} \tag{26}$$

$$= a_1b_1 + a_2b_2 + a_3b_3. \tag{30}$$

$$\mathbf{a}^2 = a^2 = a_1^{\,2} + a_2^{\,2} + a_3^{\,2}.$$

$$\mathbf{a}_1\cdot\mathbf{b}_1 = \cos(\mathbf{ab}).$$

$$(\mathbf{a}_1)^2 = \mathbf{a}_1\cdot\mathbf{a}_1 = 1.$$

If $\quad \mathbf{a}\cdot\mathbf{b} = 0, \quad$ then $\quad \mathbf{a} \perp \mathbf{b}. \qquad (27)$

$$(\mathbf{a} + \mathbf{b})\cdot(\mathbf{c} + \mathbf{d}) = \mathbf{a}\cdot\mathbf{c} + \mathbf{a}\cdot\mathbf{d} + \mathbf{b}\cdot\mathbf{c} + \mathbf{b}\cdot\mathbf{d}. \qquad (28)$$

No attention need be paid to the order of the factors.

$$\mathbf{a}\times\mathbf{b} = \boldsymbol{\epsilon}\, ab \sin(\mathbf{ab}) = -\mathbf{b}\times\mathbf{a} \qquad (33)$$

$$= \mathbf{i}(a_2b_3 - a_3b_2) + \mathbf{j}(a_3b_1 - a_1b_3) + \mathbf{k}(a_1b_2 + a_2b_1) \qquad (39)$$

$$= \begin{vmatrix} \mathbf{i} & \mathbf{j} & \mathbf{k} \\ a_1 & a_2 & a_3 \\ b_1 & b_2 & b_3 \end{vmatrix}. \qquad (40)$$

$\boldsymbol{\epsilon}$ is a unit vector, perpendicular to the plane of **a** and **b** and pointing in such a sense that as **a** is turned towards **b** a cork-screw would advance along $\boldsymbol{\epsilon}$.

$$(\mathbf{a}_1\times\mathbf{b}_1)_0 = \sin(\mathbf{ab}).$$

$$\mathbf{a}\times\mathbf{a} = 0. \qquad (34)$$

If $\quad \mathbf{a}\times\mathbf{b} = 0, \quad$ then **b** is parallel to **a**.

$$\mathbf{i}^2 = \mathbf{j}^2 = \mathbf{k}^2 = 1. \qquad \mathbf{i}\cdot\mathbf{j} = \mathbf{j}\cdot\mathbf{k} = \mathbf{k}\cdot\mathbf{i} = 0. \qquad (29)$$

$$\mathbf{i}\times\mathbf{i} = \mathbf{j}\times\mathbf{j} = \mathbf{k}\times\mathbf{k} = 0. \qquad \mathbf{i}\times\mathbf{j} = \mathbf{k}, \quad \mathbf{j}\times\mathbf{k} = \mathbf{i}, \quad \mathbf{k}\times\mathbf{i} = \mathbf{j}. \qquad (35)$$

If $\mathbf{a}'$ is the component of **a** normal to **b**, then

$$\mathbf{a}\times\mathbf{b} = \mathbf{a}'\times\mathbf{b}. \qquad (36)$$

$$(\mathbf{a} + \mathbf{b})\times(\mathbf{c} + \mathbf{d}) = \mathbf{a}\times\mathbf{c} + \mathbf{a}\times\mathbf{d} + \mathbf{b}\times\mathbf{c} + \mathbf{b}\times\mathbf{c}. \qquad (38)$$

Great attention must be paid to the order of the factors.

Products of Three Vectors.

The scalar

$$\mathbf{a}\cdot(\mathbf{b}\times\mathbf{c}) = (\mathbf{a}\times\mathbf{b})\cdot\mathbf{c} = \begin{vmatrix} a_1 & a_2 & a_3 \\ b_1 & b_2 & b_3 \\ c_1 & c_2 & c_3 \end{vmatrix} = [\mathbf{abc}]^* \qquad (49)$$

is equal to the volume of the parallelopiped of which **a, b,** and **c** are the three determining edges.

* [**abc**] represents *all* arrangements of the triple scalar products, (48), having the same cyclical order of factors.

The vector $\mathbf{a}\times(\mathbf{b}\times\mathbf{c}) = \mathbf{b}\ \mathbf{a}\cdot\mathbf{c} - \mathbf{c}\ \mathbf{a}\cdot\mathbf{b}.$ (55)

Any vector r may be represented in terms of three others by the formula

$$\mathbf{r}\,[\mathbf{abc}] = [\mathbf{rbc}]\,\mathbf{a} + [\mathbf{rca}]\,\mathbf{b} + [\mathbf{rab}]\,\mathbf{c}. \qquad (61)$$

The plane normal to **a** and passing through the terminus of **b** is

$$\mathbf{a}\cdot(\mathbf{r} - \mathbf{b}) = 0. \qquad (64)$$

The perpendicular from the origin to this plane is

$$\mathbf{p} = \mathbf{a}^{-1}\,\mathbf{a}\cdot\mathbf{b}. \qquad (68a)$$

The plane parallel to **c** and **d** through the end of **b** is

$$(\mathbf{c}\times\mathbf{d})\cdot(\mathbf{r} - \mathbf{b}) = 0. \qquad (70)$$

The plane through the three points, **a**, **b**, and **c** is

$$(\mathbf{r} - \mathbf{a})\cdot(\mathbf{a} - \mathbf{b})\times(\mathbf{b} - \mathbf{c}) = 0, \qquad (67)$$

or $\quad \boldsymbol{\phi}\cdot(\mathbf{r} - \mathbf{a}) = 0, \quad$ where $\boldsymbol{\phi} \equiv (\mathbf{a}\times\mathbf{b} + \mathbf{b}\times\mathbf{c} + \mathbf{c}\times\mathbf{a}).$

The perpendicular from the origin to this plane is

$$\mathbf{p} = \boldsymbol{\phi}^{-1}\,\boldsymbol{\phi}\cdot\mathbf{a}. \qquad (68b)$$

The line through the end of **b** and parallel to **a** is

$$\mathbf{a}\times(\mathbf{r} - \mathbf{b}) = 0. \qquad (69)$$

The equation of the sphere (or **circle**) of radius **a** with center at the origin is

$$\mathbf{r}^2 = \mathbf{a}^2. \qquad (71)$$

If the origin is at the point **c**, it becomes

$$\mathbf{r}^2 - 2\,\mathbf{r}\cdot\mathbf{c} = \mathbf{a}^2 - \mathbf{c}^2 = \text{const.} \qquad (72)$$

If the origin lies on the circumference,

$$\mathbf{r}^2 - 2\,\mathbf{r}\cdot\mathbf{a} = 0. \qquad (73)$$

DIFFERENTIATION OF VECTORS.

$$\frac{d^n\mathbf{a}}{dt^n} = \mathbf{i}\,\frac{d^n a_1}{dt^n} + \mathbf{j}\,\frac{d^n a_2}{dt^n} + \mathbf{k}\,\frac{d^n a_3}{dt^n}. \tag{77}$$

If $p \equiv \frac{d}{dt}$ and $p^n \equiv \frac{d^n}{dt^n}$, this may be written

$$p^n\mathbf{a} = \mathbf{i}\,p^n a_1 + \mathbf{j}\,p^n a_2 + \mathbf{k}\,p^n a_3, \tag{78}$$

$$p(\mathbf{a}\cdot\mathbf{b}) = p\,\mathbf{a}\cdot\mathbf{b} + \mathbf{a}\cdot p\,\mathbf{b}. \tag{79}$$

No attention need be paid to the order of the factors in a scalar product.

$$p(\mathbf{a}\times\mathbf{b}) = p\,\mathbf{a}\times\mathbf{b} + \mathbf{a}\times p\,\mathbf{b}. \tag{80}$$

Great attention must be paid to the order of the factors in a vector product.

$$\begin{aligned} p\,[\mathbf{a}\cdot\mathbf{b}\times\mathbf{c}] &= p\,\mathbf{a}\cdot\mathbf{b}\times\mathbf{c} \quad + \mathbf{a}\cdot p\,\mathbf{b}\times\mathbf{c} \quad + \mathbf{a}\cdot\mathbf{b}\times p\,\mathbf{c}, \\ p\,[\mathbf{a}\times(\mathbf{b}\times\mathbf{c})] &= p\,\mathbf{a}\times(\mathbf{b}\times\mathbf{c}) + \mathbf{a}\times(p\,\mathbf{b}\times\mathbf{c}) + \mathbf{a}\times(\mathbf{b}\times p\,\mathbf{c}). \end{aligned} \tag{81}$$

The Operator ∇ (del).

$$\nabla = \mathbf{i}\,\frac{\partial}{\partial x} + \mathbf{j}\,\frac{\partial}{\partial y} + \mathbf{k}\,\frac{\partial}{\partial z} = \text{del}, \tag{102}$$

$$d\mathbf{r} = \mathbf{i}\,dx + \mathbf{j}\,dy + \mathbf{k}\,dz,$$

$$d\mathbf{r}\cdot\nabla = \frac{\partial}{\partial x}\,dx + \frac{\partial}{\partial y}\,dy + \frac{\partial}{\partial z}\,dz = d(\ ), \tag{114}$$

$$\nabla V = \mathbf{i}\,\frac{\partial V}{\partial x} + \mathbf{j}\,\frac{\partial V}{\partial y} + \mathbf{k}\,\frac{\partial V}{\partial z} = \text{grad } V \text{ (a vector)}, \tag{106}$$

$$(\nabla V)_0 = \frac{dV}{dn},$$

$$\nabla r = \mathbf{r}_1,$$

$$\nabla\frac{1}{r} = -\frac{\mathbf{r}_1}{r^2}, \tag{109}$$

$$\nabla r^n = nr^{n-1}\nabla\mathbf{r} = nr^{n-1}\mathbf{r}_1 = nr^{n-2}\mathbf{r}, \tag{108}$$

$$\nabla(\mathbf{a}\cdot\mathbf{b}) = \nabla_a(\mathbf{a}\cdot\mathbf{b}) + \nabla_b(\mathbf{a}\cdot\mathbf{b}) \equiv \nabla(\mathbf{a}\cdot\mathbf{b})_b + \nabla(\mathbf{a}\cdot\mathbf{b})_a, \quad (110)$$

$$(\mathbf{s}_1\cdot\nabla)\,V = \mathbf{s}_1\cdot(\nabla V) = \frac{dV}{ds} = s_1\frac{\partial V}{\partial x} + s_2\frac{\partial V}{\partial y} + s_3\frac{\partial V}{\partial z}, \quad (112)$$

$$(\mathbf{s}_1\cdot\nabla)\,F = \mathbf{s}_1\cdot(\nabla F), \quad (113)$$

$$(\mathbf{s}_1\cdot\nabla)\mathbf{F} = (\mathbf{s}_1\cdot\nabla)(\mathbf{i}\,F_1 + \mathbf{j}\,F_2 + \mathbf{k}\,F_3), \quad (116)$$

$$\mathbf{a}\cdot(\mathbf{b}\cdot\nabla_F\mathbf{F}) = \mathbf{b}\cdot\nabla_F(\mathbf{a}\cdot\mathbf{F}), \quad (117)$$

$$\nabla\cdot\mathbf{F} = \frac{\partial F_1}{\partial x} + \frac{\partial F_2}{\partial y} + \frac{\partial F_3}{\partial z} = \text{div}\,\mathbf{F}\ (\text{a scalar}). \quad (119)$$

The Divergence Theorem,

$$\iint_S \mathbf{n}\cdot\mathbf{q}\,dS = \iiint_{\text{vol.}\,S} \nabla\cdot\mathbf{q}\,dv, \quad (121)$$

where $\mathbf{q}$ is any vector function and $\mathbf{n}$ is the externally-drawn unit normal to dS.

$$\nabla\times\mathbf{F} = \begin{vmatrix} \mathbf{i} & \mathbf{j} & \mathbf{k} \\ \frac{\partial}{\partial x} & \frac{\partial}{\partial y} & \frac{\partial}{\partial z} \\ F_1 & F_2 & F_3 \end{vmatrix} = \text{curl}\,\mathbf{F} \quad (125)$$

$$= \mathbf{i}\left(\frac{\partial F_3}{\partial y} - \frac{\partial F_2}{\partial z}\right) + \mathbf{j}\left(\frac{\partial F_1}{\partial z} - \frac{\partial F_3}{\partial x}\right) + \mathbf{k}\left(\frac{\partial F_2}{\partial x} - \frac{\partial F_1}{\partial y}\right).$$

$$\nabla\times(\boldsymbol{\omega}\times\mathbf{r}) = 2\,\boldsymbol{\omega},\ \text{if}\ \boldsymbol{\omega} = \text{const. vector}. \quad (126)$$

General Differentiating Formulæ for del.

$$\begin{aligned} \nabla\,(u + v) &= \nabla u + \nabla v, \\ \nabla\cdot(\mathbf{u} + \mathbf{v}) &= \nabla\cdot\mathbf{u} + \nabla\cdot\mathbf{v}, \\ \nabla\times(\mathbf{u} + \mathbf{v}) &= \nabla\times\mathbf{u} + \nabla\times\mathbf{v}. \end{aligned} \quad (127)$$

$$\begin{aligned} \nabla\,(uv) &= v\nabla u + u\nabla v, \\ \nabla\cdot(u\mathbf{v}) &= \nabla u\cdot\mathbf{v} + u\,\nabla\cdot\mathbf{v}, \\ \nabla\times(u\mathbf{v}) &= \nabla u\times\mathbf{v} + u\,\nabla\times\mathbf{v}. \end{aligned} \quad (128)$$

$$\nabla(\mathbf{u}\cdot\mathbf{v}) = \mathbf{u}\cdot\nabla\mathbf{v} + \mathbf{u}\times(\nabla\times\mathbf{v}) + \mathbf{v}\cdot\nabla\mathbf{u} + \mathbf{v}\times(\nabla\times\mathbf{u}). \qquad (129)$$

$$\nabla\cdot(\mathbf{u}\times\mathbf{v}) = \mathbf{v}\cdot\nabla\times\mathbf{u} - \mathbf{u}\cdot\nabla\times\mathbf{v}, \qquad (130)$$

$$\nabla\times(\mathbf{u}\times\mathbf{v}) = \mathbf{u}(\nabla_{uv}\cdot\mathbf{v}) - \mathbf{v}(\nabla_{uv}\cdot\mathbf{u})$$

$$= \mathbf{u}\nabla\cdot\mathbf{v} + \mathbf{v}\cdot\nabla\mathbf{u} - \mathbf{v}\nabla\cdot\mathbf{u} - \mathbf{u}\cdot\nabla\mathbf{v}. \qquad (131)$$

$$\mathbf{q}(\mathbf{r} + d\mathbf{r}) = \mathbf{q}(\mathbf{r}) + \nabla_q(d\mathbf{r}\cdot\mathbf{q}) + (\nabla\times\mathbf{q})\times d\mathbf{r}. \qquad (135)$$

$$\nabla(\mathbf{a}\cdot\mathbf{r}) = \mathbf{a}, \qquad \mathbf{a} = \text{const. vector.}$$

$$(\boldsymbol{\omega}\cdot\nabla_r)\mathbf{r} = \boldsymbol{\omega}, \qquad (118)$$

$$\nabla\cdot\mathbf{r} = 3, \qquad (123)$$

$$\nabla\cdot\mathbf{r}_1 = \frac{2}{r}.$$

$$\nabla\cdot\frac{1}{\mathbf{r}} = -\frac{1}{r^2},$$

$$\mathbf{a}\cdot\nabla\frac{1}{r} = \frac{\mathbf{a}\cdot\mathbf{r}}{r^3} = \frac{\mathbf{a}\cdot\mathbf{r}_1}{r^2},$$

$$\mathbf{b}\cdot\nabla\left(\mathbf{a}\cdot\nabla\frac{1}{r}\right) = \frac{3\,\mathbf{a}\cdot\mathbf{r}\;\mathbf{b}\cdot\mathbf{r}}{r^5} + \frac{\mathbf{a}\cdot\mathbf{b}}{r^3}, \quad \mathbf{a} \text{ and } \mathbf{b} \text{ const. vectors.}$$

$$\nabla^2\mathbf{f}(\mathbf{r}) = \mathbf{f}''(\mathbf{r}) + \frac{2\,\mathbf{f}'(\mathbf{r})}{r}.$$

Taylor's Theorem.

$$f(\mathbf{r} + \boldsymbol{\epsilon}) = e^{\boldsymbol{\epsilon}\cdot\nabla}f(\mathbf{r}). \qquad (141)$$

Stokes' Theorem.

$$\int_{\mathfrak{S}} \mathrm{F}\cdot d\mathbf{r} = \iint_{\text{cap}} \mathbf{n}\cdot(\nabla\times\mathrm{F})dS. \qquad (136)$$

$$\nabla\cdot\nabla V \equiv \nabla^2 V = \text{div grad } V = \text{del square } V \text{ (a scalar)}$$

$$= \frac{\partial^2 V}{\partial x^2} + \frac{\partial^2 V}{\partial y^2} + \frac{\partial^2 V}{\partial z^2}. \qquad (147)$$

∇^2 is called the Laplacian operator.

$$\nabla^2\mathbf{F} = \nabla^2(\mathbf{i}\,F_1 + \mathbf{j}\,F_2 + \mathbf{k}\,F_3) = \mathbf{i}\,\nabla^2 F_1 + \mathbf{j}\,\nabla^2 F_2 + \mathbf{k}\,\nabla^2 F_3$$

$$= \frac{\partial^2\mathbf{F}}{\partial x^2} + \frac{\partial^2\mathbf{F}}{\partial y^2} + \frac{\partial^2\mathbf{F}}{\partial z^2}.$$

$$\begin{aligned}
\nabla\times\nabla V &= \text{curl grad } V \equiv 0,\\
\nabla\cdot\nabla\times\mathbf{F} &= \text{div curl } \mathbf{F} \equiv 0,\\
\nabla\times(\nabla\times\mathbf{F}) &= \text{curl}^2\, \mathbf{F} = \nabla(\nabla\cdot\mathbf{F}) - \mathbf{F}(\nabla\cdot\nabla) \qquad (145)\\
&= \text{grad div } \mathbf{F} - \nabla^2\mathbf{F},\\
\nabla^2 r^m &= m(m+1)r^{m-2}, \qquad (148)\\
\nabla^2 \frac{1}{r} &= 0.
\end{aligned}$$

Gauss' Integral.

$$\iint_S \frac{\mathbf{n}\cdot\mathbf{r}}{r^2}\, dS = 4\pi \quad \text{or} = 0 \qquad (150)$$

according as the origin is taken inside or outside of the closed surface S.

Laplace's Equation.

$$\nabla^2 V = 0. \qquad (157)$$

Poisson's Equation. According to system of units chosen, it is

$$\nabla^2 V = -4\pi\rho \qquad \text{or} \qquad \nabla^2 V = -\rho. \qquad (156)$$

Its solution is

$$V = \frac{1}{4\pi}\iiint_\infty \frac{\rho\, dv}{r} \quad \text{or} \quad V = \iiint_\infty \frac{\rho\, dv}{r}, \text{ respt.} \qquad (167)$$

If V satisfies the equations

$$\nabla^2 V = 0$$

and

$$\mathbf{r}\cdot\nabla V = nV,$$

it is a **spherical harmonic** of degree n.

Green's Theorems.

$$\iint_S \mathbf{n}\cdot U\nabla V dS = \iiint_{\text{vol}} U\nabla^2 V dv + \iiint_{\text{vol}} (\nabla U\cdot\nabla V) dv, \qquad (158)$$

$$\iint_S \mathbf{n}\cdot(U\nabla V - V\nabla U)\, dS = \iiint_{\text{vol}} (U\nabla^2 V - V\nabla^2 U)\, dv. \qquad (159)$$

Green's Formula.

$$V_0 = -\frac{1}{4\pi}\iiint_{\text{Region}} \frac{\nabla^2 V}{r}\, dv$$

$$+\frac{1}{4\pi}\iint_{\text{Surfaces}} \mathbf{n}\cdot\left(\frac{1}{r}\nabla V - V\nabla\frac{1}{r}\right) dS, \tag{161}$$

$$\text{Pot } V \equiv \iiint_\infty \frac{V\,dv}{r}. \tag{168}$$

$$-\frac{1}{4\pi} \text{ pot ( ) is the inverse operator to } \nabla^2 \text{ ( )}. \tag{169}$$

pot curl ( ) = curl pot ( ), for a vector function.
∇ pot ( ) = pot ∇ ( ), for a scalar function.

Theorem of Helmholtz.

$$\mathbf{W} = -\frac{1}{4\pi}\nabla \iiint \frac{\nabla\cdot\mathbf{W}}{r} dv + \frac{1}{4\pi}\nabla\times\iiint\frac{\nabla\times\mathbf{W}}{r}\, dv \tag{175}$$

$$= -\frac{1}{4\pi}\nabla \text{ pot }(\nabla\cdot\mathbf{W}) + \frac{1}{4\pi}\nabla\times \text{ pot}(\nabla\times\mathbf{W}). \tag{178}$$

$$\int_{\mathfrak{O}} \mathbf{q}\times d\mathbf{r} = \iint_{\text{cap.}} [(\nabla\cdot\mathbf{q})\mathbf{n} - \nabla(\mathbf{q}\cdot\mathbf{n})]\, dS \tag{201}$$

$\mathbf{q}$ = any vector function.

Maxwell's Electro-magnetic Equations for Media at Rest.

$$\frac{\partial \mathfrak{F}}{\partial t} = \nabla\times\mathbf{H} \qquad \mathfrak{F} = \varepsilon\mathbf{F},$$
$$-\frac{\partial \mathfrak{H}}{\partial t} = \nabla\times\mathbf{F} \qquad \mathfrak{H} = \mu\mathbf{H}. \tag{192}$$

Linear Vector Function.

If $\boldsymbol{\phi}$ is a linear vector function,

$$\begin{aligned} \boldsymbol{\phi}(\boldsymbol{\tau} \pm \boldsymbol{\sigma}) &= \boldsymbol{\phi\tau} \pm \boldsymbol{\phi\sigma}, \\ \boldsymbol{\phi}\, a\, \boldsymbol{\tau} &= a\, \boldsymbol{\phi\tau}, \\ d(\boldsymbol{\phi\tau}) &= \boldsymbol{\phi}\, d\boldsymbol{\tau}, \\ \boldsymbol{\tau}\cdot\boldsymbol{\phi\sigma} &= \boldsymbol{\sigma}\cdot\boldsymbol{\phi}'\boldsymbol{\tau}. \end{aligned} \tag{223}$$

$\boldsymbol{\phi}'$ is said to be the conjugate function to $\boldsymbol{\phi}$.

If
$$\phi' = \phi,$$
then ϕ is said to be self-conjugate.

If
$$\boldsymbol{\omega} = \omega_1 \mathbf{i} + \omega_2 \mathbf{j} + \omega_3 \mathbf{k}$$
and
$$\phi\boldsymbol{\omega} = A\omega_1 \mathbf{i} + B\omega_2 \mathbf{j} + C\omega_3 \mathbf{k}, \tag{232}$$
then
$$\phi(\phi\boldsymbol{\omega}) \equiv \phi^2\boldsymbol{\omega} = A^2\omega_1 \mathbf{i} + B^2\omega_2 \mathbf{j} + C^2\omega_3 \mathbf{k},$$
and
$$\phi^n\boldsymbol{\omega} = A^n\omega_1 \mathbf{i} + B^n\omega_2 \mathbf{j} + C^n\omega_3 \mathbf{k}.$$
If
$$\phi^{-1}(\phi\boldsymbol{\omega}) \equiv \boldsymbol{\omega}, \tag{233}$$
then
$$\phi^{-1}\boldsymbol{\omega} = A^{-1}\omega_1 \mathbf{i} + B^{-1}\omega_2 \mathbf{j} + C^{-1}\omega_3 \mathbf{k}.$$

Coriolis' Equation. A vector **a** in fixed space, when referred to a moving space which has an angular velocity of rotation $\boldsymbol{\omega}$, satisfies the following relation:

$$\left(\frac{d\mathbf{q}}{dt}\right)_{fs} = \left(\frac{d\mathbf{q}}{dt}\right)_{ms} + \boldsymbol{\omega}\times\mathbf{q}. \tag{239}$$

D'Alembert's Equation.

$$\sum\left(m\frac{d^2\mathbf{r}}{dt^2} - \mathbf{F}\right)\cdot\delta\mathbf{r} = 0. \tag{201}$$

Euler's Equation of Motion of a Rigid Body about a Fixed Point.

$$\frac{d\mathbf{H}}{dt} + \boldsymbol{\omega}\times\mathbf{H} = \mathbf{M}. \tag{243}$$

Hamilton's Principle.

$$\delta\int_{t_1}^{t_2}(T - W)\,dt = 0. \tag{255}$$

Lagrange's Equations of Motion.

$$\frac{d}{dt}\nabla_n{}'L - \nabla_n L = 0, \tag{266}$$

or

$$\overline{\nabla}L = 0, \tag{267}$$

where

$$\overline{\nabla} \equiv \left(\frac{d}{dt}\nabla_n{}' - \nabla_n\right).$$

Hydrodynamics.

Equation of Continuity.

$$\frac{\partial \rho}{\partial t} + \nabla\cdot(\rho\,\mathbf{q}) = 0, \tag{269}$$

or

$$\frac{D\rho}{Dt} + \rho\nabla\cdot\mathbf{q} = 0. \tag{271}$$

Euler's Equation of Motion of a Fluid.

$$\frac{D(\rho\,\mathbf{q})}{Dt} = \frac{\partial(\rho\,\mathbf{q})}{\partial t} + \mathbf{q}\cdot\nabla(\rho\,\mathbf{q}) = \rho\,\mathbf{F} - \nabla p. \tag{275}$$

Circulation along the path AB.

$$\phi_{AB} = \int_A^B \mathbf{q}\cdot d\mathbf{r}.$$

ADDITIONS TO APPENDIX.

Note to § 4.

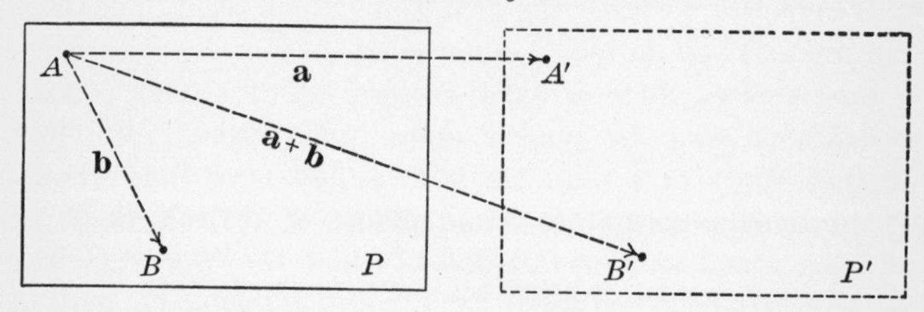

Note on Different Varieties of Vectors. — Consider a particle on the movable platform P. The particle is initially at A. If the particle remains at rest on the platform while the platform is displaced uniformly to a new position P', the particle will describe the path AA' *relatively to the ground.* This motion can be conveniently described by the vector $\overrightarrow{AA'}$ or more concisely by the single symbol **a.**

If the platform remains at rest and the particle moves uniformly to B, the path thus discribed relatively to the ground is $\overrightarrow{AB}$, or more shortly **b.**

Evidently, if the particle moves uniformly from A to B while the platform moves uniformly from P to P', the particle will finally end up at B', and will have described the path $\overrightarrow{AB}$ uniformly with respect to the ground, and in the same time. This path is **defined** to be the sum of the displacements (or vectors) **a** and **b** and is written **a + b.**

(In this case also these displacements may take place consecutively in either order and the *final* position of the particle will be the same.)

Now **it is a fact** that forces, velocities and many other physical quantities obey this same law. Hence they will obey the consequences of a calculus which follows from the above, and other definitions consistent with the facts.

There is nothing else but **convenience** which obliges us to define the sum of two vectors in the above fashion. We could have defined the sum in other ways, and by other non-contradictory definitions obtained a consistent analysis. We choose, of course, the definitions which seem to us most natural and best suited to our needs.

Free Vectors, Slide or Axial Vectors. — Quite often certain restrictions *must* be placed upon our vectors. We may restrict them to a plane in which they may slide about, or to remain normal to some plane, or restrict them to slide back and forth in the lines of which they are segments, or even attach them to a fixed point allowing no motion whatsoever.

For example: It is well known that couples may be represented by vectors perpendicular to their plane and that the effect of any couple is the same as long as its representative vector remains parallel to itself, however otherwise displaced. This is the freest kind of a vector and may be called a **Free Vector.**

On the other hand, forces produce the same effect provided only that they are not displaced out of their line of action. This kind of a vector with restricted freedom is called a **Slide Vector,** or **Axial Vector.**

Nevertheless, if we disregard for the time the known effect of displacing a force out of its line of action, *i.e.*, changing its moment about any given axis, we may consider forces as free vectors. For instance, in statics, one of the conditions of equilibrium is that the resultant of all the forces, *considered acting at a common origin,* shall vanish. Here we consider the forces as free vectors for the time being. The restriction to their line of action is intrinsically contained in the remaining rule for the vanishing of the resultant moment.

Simultaneous angular velocities compound vectorially, while finite rotations do not. Thus a rotation of θ about an axis followed by a rotation of ϕ about another axis is not, in general, equal to those rotations taken in reverse order. And yet a finite rotation has an axis, which is a direction and an amount or magnitude and is, in that sense, a vector, but it does not obey the laws of our vector analysis.

However if these rotations were to take place simultaneously the resultant rotation would be correctly found by vector addition as defined above.

It is thus seen that we are endeavoring to deduce a calculus which coincides as nearly as possible with the fundamental properties of the majority of quantities to which we apply it.

Having thus constructed a consistent analysis coinciding as closely as may be with the facts, it can of course be taught abstractly without reference to them. Later on when physical quantities are *shown* to obey the same laws as vectors have been *defined* to have, we may employ the results of the vector calculus to them without further ado.

Personally the writer does not approve of the teaching of Vector Analysis as an abstract science, nor even as a mathematical subject unless by a teacher who is thoroughly familiar with the physical results to which it applies and for which it was designed.

The vector analysis as deduced in this book is that of free vectors.

Note to § 38.

Normal, Normal Plane, Principal Normal, Binormal, Rectifying Plane.

Every line perpendicular to the tangent to a curve at the point of tangency, M, is called a Normal. These normals lie in a plane called the Normal Plane and from them two are singled out for special mention: the normal which lies in the Osculating Plane called the Principal Normal, and the normal perpendicular to this plane called the Binormal.

The plane passing through the point of tangency perpendicular to the principal normal is called the Rectifying Plane.

Thus the tangent, the principal normal and the binormal form a rectangular system of vectors. Taking the point of contact as origin, the directions of these vectors may always be taken so as to form a right-handed system. The principal normal, however, is always chosen so as to point towards the center of curvature of the curve.

Let R be the magnitude of the radius of curvature and let $\boldsymbol{\tau}$, $\boldsymbol{\nu}$, $\boldsymbol{\beta}$ be *unit* vectors along the tangent, principal normal and binormal respectively. Then

$$\boldsymbol{\tau} = \frac{d\mathbf{r}}{ds} = \mathbf{r}'. \qquad (a)$$

The vector curvature $\mathbf{c}$ ($= \boldsymbol{\rho}^{-1}$ where $\boldsymbol{\rho}$ = vector radius of curvature) is, by definition, the rate of change of the unit tangent per unit arc or

$$\mathbf{c} = \frac{d\boldsymbol{\tau}}{ds} = \boldsymbol{\tau}' = \mathbf{r}'' = \boldsymbol{\rho}^{-1}.$$

And as its direction is along the principal normal we may write it, where R is the magnitude of the radius of curvature,

$$\boldsymbol{\tau}' = \mathbf{c} = \frac{\boldsymbol{\nu}}{R} = \mathbf{r}''. \qquad (b)$$

Hence

$$\boldsymbol{\nu} = R\mathbf{r}''.$$

As $\boldsymbol{\beta}$ is a unit vector $\perp$ both to $\boldsymbol{\tau}$ and $\boldsymbol{\nu}$, it is

$$\boldsymbol{\beta} = \boldsymbol{\tau} \times \boldsymbol{\nu} = R\, \mathbf{r}' \times \mathbf{r}''. \qquad (c)$$

The direction cosines of these lines are the coefficients of $\mathbf{i}$, $\mathbf{j}$, $\mathbf{k}$ in the following equations.

$$\boldsymbol{\tau} = \mathbf{r}' = x'\mathbf{i} + y'\mathbf{j} + z'\mathbf{k},$$

$$\boldsymbol{\nu} = R\mathbf{r}'' = Rx''\mathbf{i} + Ry''\mathbf{j} + Rz''\mathbf{k},$$

$$\boldsymbol{\beta} = \boldsymbol{\tau} \times \boldsymbol{\nu} = R \begin{vmatrix} \mathbf{i} & \mathbf{j} & \mathbf{k} \\ x' & y' & z' \\ x'' & y'' & z'' \end{vmatrix}$$

$$= R \begin{vmatrix} y' & z' \\ y'' & z'' \end{vmatrix} \mathbf{i} + R \begin{vmatrix} z' & x' \\ z'' & x'' \end{vmatrix} \mathbf{j} + R \begin{vmatrix} x' & y' \\ x'' & y'' \end{vmatrix} \mathbf{k}.$$

Torsion or **Tortuosity** is defined on page 78 as

$$\mathbf{T} = \frac{d\boldsymbol{\beta}}{ds} = \boldsymbol{\beta}'. \qquad (d)$$

(Here $\boldsymbol{\beta}$ is the $\mathbf{n}$ of page 78.)

The reciprocal of the torsion is a vector called the radius of torsion $\mathbf{S}$. Hence

$$\mathbf{T} = \mathbf{S}^{-1}.$$

Formulæ of Frenet for a Space Curve.

In the investigation of the properties of space curves certain formulæ due to Frenet are of fundamental importance. They enable us to express the *derivatives* of the unit vectors along the tangent, principal normal and binormal in terms of these vectors themselves.

Differentiate (a) giving

$$\boldsymbol{\tau}' = \mathbf{r}'',$$

which by (b) becomes

$$\boldsymbol{\tau}' = \frac{\boldsymbol{\nu}}{R}. \qquad \text{I}$$

Differentiate (c) giving

$$\boldsymbol{\beta}' = \boldsymbol{\tau}' \times \boldsymbol{\nu} + \boldsymbol{\tau} \times \boldsymbol{\nu}',$$

which by I becomes

$$\boldsymbol{\beta}' = \frac{\boldsymbol{\nu} \times \boldsymbol{\nu}}{R} + \boldsymbol{\tau} \times \boldsymbol{\nu}' = \boldsymbol{\tau} \times \boldsymbol{\nu}'.$$

From this equation we see that $\boldsymbol{\beta}'$ is $\perp$ to $\boldsymbol{\tau}$; and since $\boldsymbol{\beta}$ is a unit vector $\boldsymbol{\beta}'$ is $\perp$ to $\boldsymbol{\beta}$; it is therefore parallel to $\boldsymbol{\nu}$ and we may write

$$\boldsymbol{\beta}' = \frac{\boldsymbol{\nu}}{S}, \qquad \text{II}$$

where S is a scalar whose absolute value, as $\boldsymbol{\nu}$ is a unit vector, is the radius of torsion, by definition, equation (d).

Again, since $\boldsymbol{\nu}$ is at right angles to $\boldsymbol{\beta}$ and $\boldsymbol{\tau}$, we may write

$$\boldsymbol{\nu} = \boldsymbol{\beta} \times \boldsymbol{\tau},$$

which equation on differentiating becomes

$$\boldsymbol{\nu}' = \boldsymbol{\beta}' \times \boldsymbol{\tau} + \boldsymbol{\beta} \times \boldsymbol{\tau}'.$$

Using II and I we obtain

$$\boldsymbol{\nu}' = \frac{\boldsymbol{\nu} \times \boldsymbol{\tau}}{S} + \frac{\boldsymbol{\beta} \times \boldsymbol{\nu}}{R}$$

$$= -\frac{\boldsymbol{\beta}}{S} - \frac{\boldsymbol{\tau}}{R}. \qquad \text{III}$$

Equations I, II, III are Frenet's equations; they express the first derivatives of $\boldsymbol{\tau}$, $\boldsymbol{\nu}$ and $\boldsymbol{\beta}$ in terms of themselves and the scalars R and S.

Formula for the Torsion. — From the preceding we can easily obtain a simple formula for the torsion of a curve. From II and (c) we have

$$\boldsymbol{\beta}' = \frac{\boldsymbol{\nu}}{S} = (\boldsymbol{\tau} \times \boldsymbol{\nu})' = (R\mathbf{r}' \times \mathbf{r}'')'$$

$$= R'\mathbf{r}' \times \mathbf{r}'' + R\mathbf{r}'' \times \mathbf{r}'' + R\mathbf{r}' \times \mathbf{r}'''.$$

The second term is identically zero. By multiplying by equation (b), $\boldsymbol{\nu} = R\mathbf{r}$,″ we have

$$\boldsymbol{\beta}' \cdot \boldsymbol{\nu} = \frac{\boldsymbol{\nu} \cdot \boldsymbol{\nu}}{S} = \frac{1}{S} = R\mathbf{r}'' \cdot [R'\mathbf{r}' \times \mathbf{r}'' + R\mathbf{r}' \times \mathbf{r}'''].$$

Hence, as the first triple scalar product vanishes

$$\frac{1}{S} = T = -R^2 [\mathbf{r}'\mathbf{r}''\mathbf{r}''']. \qquad \text{IV}$$

Exercise: Express equations I, II, III, IV in Cartesian coördinates. Show that for a plane curve the torsion vanishes, and conversely.

Path Described by an Electron in a Uniform Magnetic Field.

As an interesting application of the vector method consider the motion of an electron in a uniform magnetic field of strength $\mathbf{H}$. It is well known by the experiments of Rowland that a moving electrical charge is equivalent to a current. Let e be the value of the charge, m the mass of the electron, and $\mathbf{v}$ its vector velocity. The force acting on a linear conductor of length $d\mathbf{r}$ is

$$\mathbf{F} = i\, d\mathbf{r} \times \mathbf{H}.$$

Let this current i be due to the convection of electricity carried by the n electrons contained in the element $d\mathbf{r}$, moving with the common velocity $\mathbf{v}$, then

$$i\, d\mathbf{r} = ne\,\mathbf{v}.$$

Hence the force on a single electron is

$$F = \frac{ne\,\mathbf{v} \times \mathbf{H}}{n} = e\,\mathbf{v} \times \mathbf{H}.$$

This force produces an acceleration $\dot{\mathbf{v}}$ on the electron, so that the equation of motion is

$$m\dot{\mathbf{v}} = e\,\mathbf{v} \times \mathbf{H}.$$

Putting $\mathbf{h} \equiv \frac{e}{m}\mathbf{H}$, the differential equation of the path is

$$\dot{\mathbf{v}} = \mathbf{v} \times \mathbf{h}. \qquad (1)$$

From the equation we see that the acceleration is normal to the path, hence the *speed* is constant. As the acceleration is also normal to $\mathbf{h}$ the velocity component parallel to $\mathbf{h}$ is constant, and hence the whole acceleration is always parallel to the plane normal to $\mathbf{h}$.

As $\mathbf{h}$ is a constant vector, we have from (1)

$$\mathbf{h} \cdot \dot{\mathbf{v}} = \mathbf{h} \cdot \mathbf{v} \times \mathbf{h} = 0 = \frac{d}{dt}(\mathbf{h} \cdot \mathbf{v}), \qquad (2)$$

so that the angle the path makes with the field is constant. From (1) again, since $\mathbf{v}$ and $\mathbf{h}$ and their included angle are constant, the *magnitude* of the acceleration is constant.

The radius of curvature of any path is related to the speed and the normal acceleration by equation (92), page 81,

$$\mathbf{a} = \frac{\mathbf{v}^2}{\boldsymbol{\rho}}.$$

Hence

$$\boldsymbol{\rho} = \frac{\mathbf{v}^2}{\dot{\mathbf{v}}} = \frac{\mathbf{v}^2}{\mathbf{v}\times\mathbf{h}}. \qquad (3)$$

And since the magnitudes of $\mathbf{v}$ and $\mathbf{v}\times\mathbf{h}$ are constant so is the magnitude of $\boldsymbol{\rho}$.

$$\boldsymbol{\rho}_0 = \frac{v}{h \sin(\mathbf{vh})}. \qquad (4)$$

The component of the speed normal to $\mathbf{h}$ is

$$v_1 = v \sin(\mathbf{vh}).$$

The radius of curvature of the path in the plane normal to $\mathbf{h}$ is, similarly to (3),

$$\rho_1 = \left(\frac{\mathbf{v}_1^2}{\mathbf{v}_1\times\mathbf{h}}\right)_0 = \frac{v^2 \sin^2(\mathbf{vh})}{v \sin(\mathbf{vh})\, h \sin(\mathbf{v}_1\mathbf{h})} = \frac{v \sin(\mathbf{vh})}{h} \qquad (5)$$

as $\sin(\mathbf{v}_1\mathbf{h}) = 1$, and hence the path is a circle of radius ρ_1.

Thus the motion is completely determined. It is a curve of constant curvature, described with constant speed whose projection on a plane normal to the magnetic lines is a circle of radius ρ_1. The velocity parallel to the field is constant. It is therefore a right circular helix whose axis is parallel to the lines of force.

Comparing equations (4) and (5) we see that the radius of curvature of a curve and that of its projection on a plane are related by the formula,

$$\rho_1 = \rho \sin^2(\theta), \qquad (6)$$

where θ is the angle between the curve and the normal to the plane. This holds for every curve, because any small portion of it may be considered to be a portion of some helix. This result is due to Euler.

A circular helix is sometimes defined as a curve having (*a*) constant curvature and (*b*) constant torsion. We can also prove that the above path is a helix by proving (*b*). Equation (4) shows that the curvature is constant.

The torsion T is by IV of Frenet's formulæ

$$T = \rho^2 [\mathbf{r}' \, \mathbf{r}'' \, \mathbf{r}'''] \tag{7}$$

where the primes denote differentiation with respect to the arc s.

Now
$$\mathbf{r}' = \dot{\mathbf{r}} \frac{dt}{ds} = \frac{\mathbf{v}}{v};$$

and since $\mathbf{v}$ is constant

$$\mathbf{r}'' = \ddot{\mathbf{r}} \left(\frac{dt}{ds}\right)^2 = \frac{\dot{\mathbf{v}}}{v^2}$$

and

$$\mathbf{r}''' = \frac{\ddot{\mathbf{v}}}{v^3}.$$

Substituting in (7)

$$T = \left(\frac{v^2}{\dot{v}}\right)^2 \left[\frac{\mathbf{v}}{v} \times \frac{\dot{\mathbf{v}}}{v^2} \cdot \frac{\ddot{\mathbf{v}}}{v^3}\right] = \frac{1}{v^2} \frac{[\mathbf{v} \times (\mathbf{v} \times \mathbf{h})] \cdot \ddot{\mathbf{v}}}{(\mathbf{v} \times \mathbf{h})^2}.$$

By differentiating (1)

$$\ddot{\mathbf{v}} = \dot{\mathbf{v}} \times \mathbf{h} = (\mathbf{v} \times \mathbf{h}) \times \mathbf{h}.$$

Hence

$$T = \frac{1}{v^2} \frac{[\mathbf{v} \times (\mathbf{v} \times \mathbf{h})] \cdot [(\mathbf{v} \times \mathbf{h}) \times \mathbf{h}]}{(\mathbf{v} \times \mathbf{h})^2},$$

which reduced by (58) becomes

$$T = \pm \frac{\mathbf{v} \cdot \mathbf{h}}{v^2}.$$

As the two parts of the fraction are separately constant the torsion is constant.

The result (4) obtained above, otherwise written

$$\frac{m}{e} v = H \rho \sin (vH),$$

is of great importance in obtaining a relation between $\left(\frac{m}{e}\right)$ and v for an electron, which in combination with other relations enables us to determine their separate values.

Note to § 58.

Two Proofs of Stokes' Theorem.

1° The line integral around the bounding curve, Fig. 54, is equal to the sum of the line integrals around the elementary parallelograms into which the surface may be considered divided. For every side of these parallelograms is integrated twice and in opposite directions, and the results cancel, except the sides which coincide with the bounding curve.

Consider an elementary parallelogram $ABCD$; let $AB = d\boldsymbol{\rho}_1$ and $AD = d\boldsymbol{\rho}_2$. Let $(d\mathbf{F})_1$ be the increment of $\mathbf{F}$ along $d\boldsymbol{\rho}_1$ and $(d\mathbf{F})_2$ that along $d\boldsymbol{\rho}_2$. Then

$$(d\mathbf{F})_1 = d\boldsymbol{\rho}_1 \cdot \nabla \mathbf{F},$$
$$(d\mathbf{F})_2 = d\boldsymbol{\rho}_2 \cdot \nabla \mathbf{F}.$$

The vector point-function $\mathbf{F}$ has at the point A the value $\mathbf{F}$; at B the value $\mathbf{F} + (d\mathbf{F})_1$; at D the value $\mathbf{F} + (d\mathbf{F})_2$.

The line integral around the parallelogram is then by definition

$$\begin{aligned} &\mathbf{F} \cdot d\boldsymbol{\rho}_1 - [\mathbf{F} + (d\mathbf{F})_2] \cdot d\boldsymbol{\rho}_1 - \mathbf{F} \cdot d\boldsymbol{\rho}_2 + [\mathbf{F} + (d\mathbf{F})_1] \cdot d\boldsymbol{\rho}_2 \\ &\quad = (d\mathbf{F})_1 \cdot d\boldsymbol{\rho}_2 - (d\mathbf{F})_2 \cdot d\boldsymbol{\rho}_1, \\ &\quad = (d\boldsymbol{\rho}_1 \cdot \nabla_F)\mathbf{F} \cdot d\boldsymbol{\rho}_2 - (d\boldsymbol{\rho}_2 \cdot \nabla_F)\mathbf{F} \cdot d\boldsymbol{\rho}_1, \\ &\quad = (d\boldsymbol{\rho}_1 \times d\boldsymbol{\rho}_2) \cdot (\nabla \times \mathbf{F}), \qquad \text{(by 58)}, \\ &\quad = (\mathbf{n} \cdot \nabla \times \mathbf{F})\, dS. \end{aligned}$$

Because $d\boldsymbol{\rho}_1 \times d\boldsymbol{\rho}_2 = \mathbf{n}\, dS$ where $\mathbf{n}$ is the unit normal to this elementary area dS.

The theorem follows immediately.

2° Again by means of the Divergence Theorem which is the fundamental formula for the transformation of volume into surface integrals and vice versa, we can deduce Stokes' Theorem which is the fundamental formula for the transformation of surface into line integrals.

Apply the formula

$$\iiint \nabla \cdot \mathbf{F}\, dv = \iint \mathbf{F} \cdot \mathbf{n}\, dS, \qquad \mathbf{(1)}$$

(which is the Divergence Theorem and is to be taken throughout the volume of, and over the surface of, any closed space) to an infinitesimal right cylinder, of height h and base whose area is S. Let ds be an element of the contour of the base, and $\mathbf{c}_1$ a unit vector $\perp$ to the base.

Suppose the cylinder so small that $\mathbf{F}$ may be considered constant throughout it, so that the two surface integrals over the two plane ends cancel each other and their sum vanishes.

Replace $\mathbf{F}$ in (1) by $\mathbf{c}_1 \times \mathbf{F}$, then

$$\iiint \nabla \cdot (\mathbf{c}_1 \times \mathbf{F})\, dv = \iint \mathbf{c}_1 \times \mathbf{F} \cdot \mathbf{n}\, ds.$$

By (130) $\qquad \nabla \cdot (\mathbf{c}_1 \times \mathbf{F}) = \mathbf{F} \cdot \nabla \times \mathbf{c}_1 - \mathbf{c}_1 \cdot \nabla \times \mathbf{F}$

in which the first term of the right-hand number vanishes, as $\mathbf{c}_1$ is a constant vector; hence

$$\iiint \mathbf{c}_1 \cdot \nabla \times \mathbf{F}\, dv = -\iint \mathbf{c}_1 \times \mathbf{F} \cdot \mathbf{n}\, dS$$

$$= -\iint \mathbf{F} \cdot \mathbf{n} \times \mathbf{c}_1\, dS.* \qquad (2)$$

But $dv = h\, dS$, and $dS = h\, ds$ so that the volume and surface integrals become surface and line integrals respectively, and

$$h \iint \mathbf{c}_1 \cdot \nabla \times \mathbf{F}\, dS = h \int \mathbf{F} \cdot \mathbf{n} \times \mathbf{c}_1\, ds.$$

Now suppose the base to be an element of area of a surface bounded by a closed contour. Then $\mathbf{c}_1$ becomes a unit vector normal to the surface, which we call $\mathbf{n}$, and what was formerly $\mathbf{n} \times \mathbf{c}_1$ in (2) becomes a unit vector in the direction of $d\mathbf{r}$, so that

$$\mathbf{F} \cdot \mathbf{n} \times \mathbf{c}_1\, ds = \mathbf{F} \cdot d\mathbf{r}.$$

Summing up the elements

$$\sum \iint \mathbf{n} \cdot \nabla \times \mathbf{F}\, dS = \sum \int \mathbf{F} \cdot d\mathbf{r}.$$

The summation of the surface integrals means simply that the integral

$$\iint \mathbf{n} \cdot \nabla \times \mathbf{F}\, dS$$

is to be taken over the entire surface.

In summing up the line integrals, the contour of every elementary area is traced twice, but in opposite directions except those forming part of the contour.

* The negative sign indicates relationship of direction around the centour and the vector $\mathbf{n}$.

Hence the sum of the line integrals becomes the line integral around the bounding curve and we have again Stokes' Theorem.

Note to § 65.

Another Proof of Gauss's Theorem.

By means of the Divergence Theorem an easy proof of Gauss's theorem may be given.

The problem is to evaluate the integral

$$\iint_S \frac{\mathbf{n}\cdot\mathbf{r}_1}{\mathbf{r}_0^2}\, dS$$

over the closed surface S.

This integral may be written

$$-\iint \mathbf{n}\cdot\nabla\left(\frac{1}{r}\right) dS, \qquad \text{by (109).}$$

There are three cases, according as the origin is without, within, or on the surface S.

Case I. The origin is without. In this case $\mathbf{r}$ can never be zero, hence because by the Divergence Theorem

$$\iint \mathbf{n}\cdot\nabla \frac{1}{r}\, dS = \iiint \nabla^2 \frac{1}{r}\, dv,$$

and because

$$\nabla^2 \frac{1}{r} = 0, \qquad \text{by § 64,}$$

we have

$$\iint \frac{\mathbf{n}\cdot\mathbf{r}_1}{\mathbf{r}_0^2}\, dS = 0.$$

Case II. The origin is within. Surround the origin by a small sphere of radius ϵ; then the origin is excluded from the region bounded by S and S'. Hence the required integral taken over both surfaces is zero, by Case I, *i.e.*,

$$\iint_S \frac{\mathbf{n}\cdot\mathbf{r}_1}{\mathbf{r}_0^2}\, dS + \iint_{S'} \frac{\mathbf{n}\cdot\mathbf{r}_1}{\mathbf{r}_0^2}\, dS = 0.$$

But for the sphere, $\mathbf{r}_0 = \epsilon$, $\mathbf{n}\cdot\mathbf{r}_1 = -1$,

$$\iint_{S'} dS = 4\pi\epsilon^2;$$

$$\therefore \iint_{S'} \frac{\mathbf{n}\cdot\mathbf{r}_1}{\mathbf{r}_0^2} dS = -4\pi.$$

Hence

$$\iint_{S} \frac{\mathbf{n}\cdot\mathbf{r}_1}{\mathbf{r}_0^2} dS = 4\pi.$$

CASE III. The origin is on the surface. Exclude the point by an elementary hemisphere.

Then proceeding as in Case II, we find that the required integral is equal to minus the integral over the hemisphere, *i.e.*, to 2π.

Note to § 52.

Other Integration Theorems.

We can evaluate two volume integrals in a manner similar to that given by the Divergence Theorem. First, to find

$$\iiint \nabla F \, dv,$$

where F is a scalar point-function. Let $\mathbf{c}$ be a constant vector,

$$\iiint \mathbf{c}\cdot\nabla F \, dv = \iiint \nabla\cdot(\mathbf{c}F) \, dv.$$

But by the Divergence Theorem

$$\iiint \nabla\cdot(\mathbf{c}F) \, dv = \iint \mathbf{n}\cdot(\mathbf{c}F) \, dS,$$

$$= \mathbf{c}\cdot\iint \mathbf{n}F \, dS,$$

$$\therefore \mathbf{c}\cdot\iiint \nabla F \, dv = \mathbf{c}\cdot\iint \mathbf{n}F \, dS,$$

or since $\mathbf{c}$ is perfectly arbitrary,

$$\iiint \nabla F \, dv = \iint \mathbf{n} F \, dS.$$

Again

$$\mathbf{c} \cdot \iiint \nabla \times \mathbf{F} \, dv = \iiint \mathbf{c} \cdot \nabla \times \mathbf{F} \, dv,$$

$$= \iiint \nabla \cdot (\mathbf{F} \times \mathbf{c}) \, dv,$$

$$= \iint \mathbf{n} \cdot (\mathbf{F} \times \mathbf{c}) \, dS,$$

$$= \iint \mathbf{c} \cdot (\mathbf{n} \times \mathbf{F}) \, dS.$$

Hence finally

$$\iiint \nabla \times \mathbf{F} \, dv = \iint \mathbf{n} \times \mathbf{F} \, dS.$$

By similar processes, starting with the Divergence Theorem, which is seen to be a formula of *fundamental importance*, many other relations between surface and volume integrals, and indeed also between line and surface integrals, using the device employed in the second proof of Stokes' Theorem above, may be deduced.

INDEX.